DATE DUE FOR RETURN		

This book may b
before the ab

The Mathematical World
of Walter Noll

Springer
*Berlin
Heidelberg
New York
Barcelona
Budapest
Hong Kong
London
Milan
Paris
Santa Clara
Singapore
Tokyo*

Walter Noll

Yurie A. Ignatieff

The Mathematical World
of Walter Noll

A Scientific Biography

Springer

Dr. Yurie A. Ignatieff

Institut für Philosophie
Philipps-Universität
Wilhelm-Röpke-Strasse 6
D-35032 Marburg, Germany

The publication of this book was supported
by the Alexander von Humboldt Foundation
Bonn, Germany

With 10 Figures and 7 Tables

Library of Congress Cataloging-in-Publication Data
Ignatieff, Yurie A., 1960 –
The mathematical world of Walter Noll : a scientific biography / Yurie A. Ignatieff.
p. cm.
Includes bibliographical references (p. –) and index.
ISBN 3-540-59440-X (hardcover : alk. paper)
1. Noll, W. (Walter), 1925 – .   2. Mathematicians –United States – Biography.   I. Title.
QA29.N65I46   1996
510'.92–dc20   [B]
95-25602

ISBN 3-540-59440-X   Springer-Verlag Berlin Heidelberg New York

© Springer-Verlag Berlin Heidelberg 1996
Printed in Germany

Cover design: E. Kirchner, Heidelberg
Typesetting: Data conversion by K. Mattes, Heidelberg
SPIN 10491106   55/3144 – 5 4 3 2 1 0 – Printed on acid-free paper

Inge Lind mit Liebe

Preface

I would like to thank all those people in the life of Walter Noll who helped me to write this book: Chancellor of Indiana University Professor Herman B. Wells, Dr. Klaus André, Professor Bernard D. Coleman, Stella DeVito, Dr. Peter Fränkel, Professor Morton E. Gurtin, Studiendirektor i.R. Rudolf Hohensee, Faye Mark, Professor Victor J. Mizel, Professor Dietrich Morgenstern, Professor Ralph Raimi, Professor Juan J. Schäffer, Professor Clifford A. Truesdell III, Professor Epifanio G. Virga, Dr. Paul Winkler and many others.

I will be always indebted to Inge Lind, my teacher and friend, for a careful reading of the manuscript and her invaluable comments.

This book would never have been written without the compassion and assistance of many people in Bochum and Marburg. I am especially grateful to Ursula Arras, Dieter and Karola Behm, Achim Bühl, Birgit Berger, Professor Klaus Böhmer, Ute Hagen, Irene Joraszik, Professor Hans-Heinrich Körle, Elke and Manfred Kuhne, Jutta Küster, Johannes Lind, Inka Lins, Annemarie Matt, Astrid Milenz, Matthias Näher, Eckart Rösch, Manfred Schönsee, Professor Werner von Seelen, Christa Seip, Heike Willig, Anita Wolf and many others.

I am grateful to the editorial staff of the Springer-Verlag in Heidelberg for its excellent work on this book. My special thanks go to Professor Wolf Beiglböck, Brigitte Reichel-Mayer, Gertrud Dimler, Sabine Landgraf, Dr. Victoria Wicks and many others.

The publication of this book was supported by the Alexander von Humboldt Foundation in Bonn. I express especial gratitude for it to Dr. Dietrich Papenfuß, Professor Dietrich Morgenstern, and Sabine Döring.

Philipps University, Marburg
August 1995   *Yurie A. Ignatieff*

Contents

Introduction

This introduction is intended for historians of mathematics. I want to describe here some professional topics and ask other readers to skip it.

It was not my intention from the very beginning of the work for this book to write it for Walter Noll or the Society for Natural Philosophy. I intended to make an objective scientific-historical analysis of Walter Noll's life and mathematical activity. I was very lucky to get a copy of his autobiographical notes and to work in his archive at Carnegie-Mellon University. Fortunately, I am the first historian of mathematics to write about Walter Noll's mathematical world and can thus avoid polemics with other authors and a discussion of plagiarism.

Many of the historians of mathematics would reject the topic in principle as belonging to "contemporary mathematics" or describing a "living mathematician". They would say it could not be the subject of a "historical study". My answer to them is very short. Modern scientific-technical revolutions force the scientific personnel of our industry to take decisions on the use of this or that mathematical theory in a very short time. They want to know now – not in fifty years or so – what is available and about its quality. Historians of mathematics must play their role here, because, contrary to all other mathematical specialists, they possess the method of scientific-historical analysis. If they could take this consulting task seriously and responsibly they would make their discipline more useful for society.

I can not deny that the writing of a book about a living mathematician gives rise to certain ethical problems which many historians of mathematics would prefer to avoid. The most important of them is the question of priority. Nobody can be credited for clever avoidance in the history of mathematics, and I reject the idea that it is meaningless to give a mathematician his due. One should pose this ethical problem and solve it.

I was not credulous in respect to the recollections of Walter Noll, B.D. Coleman and others. A considerable advantage of writing about a living mathematician is that one has a sufficient number of sources of information. However, I cannot claim that my analysis is perfect. The people who gave me interviews or wrote their recollections for my book can be suspected of falsifying facts in their stories. I tried to remove doubts in all uncertain cases through a correlation with other sources. I have not tried to bend before the official German or other views of the history. It is an honorable duty and one of the main tasks of a historian of science to see the history with his own eyes and to judge it according to his cultural development.

I tried to avoid the Solomon syndrome, mentioned by Kenneth O. May, and only in several cases allowed myself to criticize the great master of mathematics, Walter Noll. Like many other mathematicians, he looks at the real world through the glasses of modern mathematics with all their advantages and disadvantages. For understandable reasons, Walter Noll did not want to tell all about his life in Nazi Germany. I had to reconstruct the missing information by deduction from established facts and by reasonable inference and hope to avoid criticism from historians of mathematics for writing a "speculative history".

From my contacts with other mathematicians in the former USSR, FRG, and USA, I know that many of them would consider the mathematical works of Walter Noll and his colleagues from the Society for Natural Philosophy as a part of engineering, mechanics, physics, and even chemistry, but not as a part of mathematics. I am very sorry for the existence of such narrow-minded views among mathematicians. As a philosopher of science, I assure historians of mathematics that it is not possible to draw a perfect boundary between mathematics and other sciences. Its existence is one of the common myths which are used to cover one's dogmatism, incompetence, and unwillingness. I would recommend my book to the students of mathematics to avoid this dangerous mathematical epidemic and to learn real mathematics from Walter Noll, one of its most prominent representatives.

1 In the Flatland

It is impossible to give a scientific definition of motherland. Partly, this is because the word concerns our private, intimate, spiritual sphere, and we dislike invaders in it. It is also because of the weakness of the analytic method of our science. It fails to split the whole into such pieces and relations among them which can, according to systems science, be united in the same whole. Since we all have different cultural backgrounds, it is a vain project even to agree on the basic words for the definition of motherland. However, if we don't have a definition we can't understand the Germans and the role of Germany in their life. It is with this in view that I have chosen E.A. Abbott's term "Flatland" to serve as an illustration, approximation, motion to the infinite boundary value, which was Germany before its culture became an object of an unprecedented reeducation campaign [1.1]. Walter Noll, who was born in Berlin on January 7, 1925, remains a German in his heart.

His father, Franz Noll, originated from a Dutch family in Rotterdam. From childhood he possessed characteristics that were very much like those of the Germans: practical cleverness, energy, initiative, and respect for knowledge. Although he emigrated to Germany in 1890 as a boy, he didn't accept Germany as his motherland and dreamed of going to the USA. Neither the eight years of German schooling in Thüringen nor his marriage to a German country girl, Martha Janssen, could change this. He kept thinking that American culture was superior to the German one. However, he studied for two years at a German trade school and then got a job at Jaroslaw's Erste Glimmerwarenfabrik in Berlin. He managed to make a career there, advancing from a common worker to a division manager [1.2]. As a member of the German Social-Democratic Party (SPD), he took an active part in the political fights of "Genossen und Genossin-

nen" [1.3]. Within the party, he belonged to the centrists and rejected armed struggle as a means to aquire state power. His wife,
who worked before marriage as a maid, was entirely unpolitical.

Walter Noll went to school for the first time in the spring of 1931.
It was a communal primary school in Berlin-Biesdorf. In the summer of 1932, Franz Noll and his family moved to their own house in
Miersdorf by Zeuthen, a suburb of Berlin [1.4]. He was unknown
there, which saved the family from political repressions after the
German fascists came to power. Franz Noll taught his only son the
basic rules of double life: to pretend to be a fellow traveller of the
ruling regime and not to talk about politics outside his own family.
Walter learned from his father that Germany was doomed to unleash
a new world war and to lose it because of the USA. Franz Noll compared the Nazis with criminals. Walter Noll could see their felonies
with his own eyes. In Biesdorf, the Nazi stormtroopers kidnapped
and beat to death the Noll family physician, Dr. Philippsthal, who
was Jewish. Another incident involved a friend of the Noll family.
Walter Noll remembered that this man made some critical remarks
about the Nazi government to a stranger in the street. As a result,
he was arrested and sentenced to several months in a concentra-
tion camp. He came back a broken man and soon committed suicide
[1.5]. After the family moved to Miersdorf, Walter Noll continued
his education at a primary school in Zeuthen. Walter Noll remembers
his first school years: "During the first four grades, my performance
in school had been only slightly above average, but my teacher as-
sured my parents that I had a good chance of success in academic
high school" [1.6].

In the spring of 1935, he received a free place at a high school for
boys in Eichwalde, district Teltow [1.7]. It was a liberal German-
nationalistic school. The teachers of Walter Noll can be divided
into two groups: formal pedagogical officials, who taught their
subjects indifferently, most of which belonged to the Nazi party
("Parteigenossen"), and sometimes wore the Nazi uniform or the
badge of the Nazi party; and the rest of the teachers [1.8]. There were
no fanatical Nazi teachers at this school, and they were not generally
in much force in Germany. Sometimes, they could not even follow
the logic of Hitler's politics and criticized him. Most of the teachers

viewed the Nazi school politics with concern, and kept to the German teaching traditions and morality. They had to be "loyal" to the Nazi regime, because they were afraid of being denounced by their pupils. If a pupil mocked the Nazis, the teacher made use of corporal punishment, which was allowed again by the Nazis, but did not shop the pupil to the school administration or to the German secret police. All the teachers received nicknames. For example, the mathematics teacher, Fuhrmeister, was called "Bommel" (bobble) [1.9].

Most of Walter Noll's classmates believed in the new Germany. Sometimes, Walter Noll allowed himself to make frivolous remarks about Nazis. Although this disloyal behavior was criticized by the others, nobody denounced him to the school administration or to the German secret police. Nobody wanted to receive the nickname "Judas" or "traitor". There was no tension between Walter Noll and his classmates who belonged to the Nazi youth organizations "Jungvolk" and "Hitlerjugend" (HJ). Even these pupils were not fanatical, passionate, ardent young Nazis. They were no more than children, attracted by the Nazi uniforms, banners, and parades. When a teacher entered the classroom, all the pupils had to salute him with a raised hand and say "Heil Hitler!". The swastika flag had been hanging in the school yard. All the pupils had to visit the "communal receptions" to listen to Hitler's speeches on the radio. While his HJ schoolmates learned comradeship and some practical things, such as camping, orienteering with compass and map, and shooting, which were necessary for the military service, Walter Noll kept himself to himself. At the high school for boys in Eichwalde, he had only one real friend, Wolfgang Otto, who was later killed in the war.

Among the school subjects, Walter Noll liked French and Latin. He remembered his progress in arithmetic: "My facility with arithmetic was never very good, and still isn't. My grades in arithmetic almost never exceeded a C. My memory for numbers is extremely poor" [1.10]. Gymnastics, swimming, and other kinds of sport were an important part of the curriculum. Training was five days a week. Walter Noll failed in all sports, so he didn't like them [1.11]. He managed finally to avoid them by obtaining a medical certificate.

In biology lessons, Walter Noll had to learn the fundamentals of Nazi ethnogeny. The school geography was full of Nazi geopolitics. Many subjects were only slightly influenced by the Nazi political doctrine: chemistry, music and drawing. In history lessons, the teacher went no further than the Bismarck era or the history of Prussia. Jewish poets and writers were avoided by the teacher of German literature. The instruction in the German language was nationalistic. The school had a good workshop, where one could learn to work with metal, paper, and wood. Some teachers had to give the so-called "national-political lessons", where pupils had to learn the Nazi political vocabulary [1.12].

In the seventh grade, Walter Noll learned the fundamentals of Euclidean geometry. However, he couldn't relate to the Nazi teacher Fuhrmeister. Walter Noll disliked him for his pedantic and authoritarian style of instruction. One day, Fuhrmeister told the class about imaginary worlds, inhabited by animated geometrical creatures, following the novel "Flatland" by E.A. Abbott. According to their dimensions, these geometrical figures inhabited three worlds: Lineland, Flatland, and Spaceland [1.13]. When the time for questions came, Walter Noll stood up and suddenly commented: "Since all matter consists of atoms and molecules, and since these are three-dimensional, there cannot be any two-dimensional objects" [1.14]. At the time of this event, Walter Noll was reading a popular book "Du und die Natur" (You and Nature), which introduced in a popular way some of the latest achievements in physics and chemistry [1.15]. For the first time, he learned from it about the curious world of the natural elements and about relativity theory. Fuhrmeister was shocked by this comment. He could not find the words to answer and just ordered Walter Noll to sit down. In the next geometry lesson, he asked Walter Noll to give an exposition of "Flatland". Walter did it perfectly and was rewarded with the highest grade, which until then had never been awarded. This mark decided the fate of Walter Noll. He decided to be a mathematician.

From the spring of 1938, Walter Noll made physics and mathematics his favourite subjects. He had the great luck to meet a teacher who discovered his mathematical talent and helped to develop it. This teacher was called Rudolf Hohensee. He was born in 1907, and both his parents were teachers. He received his university diploma on January 17, 1933, just two weeks before Hitler came to power. He wasn't a Nazi teacher and even allowed himself often to sneer at the so-called "German physics" and at the Nazi leaders. When Gregor Strasser once appealed in the press for funds for new torpedoes, Hohensee commented in class: "A torpedo costs 20 000 marks. When does your father earn 20 000 marks?" Walter Noll remembers even today Hohensee's first physics lesson: "The teacher, Rudolf Hohensee was his name, sent an electric current through a long wire, which became red hot and sagged. Then he asked: 'What do we learn from this?' He elicited from us that when things become hot, they expand. And then he said with an absolutely deadpan voice: 'This is why the days and the vacations are longer in the summer'" [1.16].

Walter Noll became Hohensee's favorite pupil. He invited Noll very often to his house on Sundays in order to teach him something new and usually beyond the ordinary school program: fundamentals of mathematical analysis, set theory, relativity theory, quantum mechanics, and other subjects. In September 1938, Rudolf Hohensee married a girl who worked as an assistant at the TU in Berlin. Liese Hohensee liked Walter Noll as her own son. Walter Noll devoted his doctoral dissertation to "his friend and teacher Rudolf Hohensee". What kind of friendship was it? Rudolf Hohensee remembers the following story. In 1939, he wanted to buy some school equipment for 6000 marks from a firm in Berlin. The money was to be paid by the community in Eichwalde. However, Germany began a war against Poland, and the community refused to pay the bill under this pretext. Rudolf Hohensee and the manager of the firm decided to force it to pay. The manager signed the bill once more with a date which was before the beginning of the war. The money was paid, and the school got its equipment. However, it was illegal, and the teacher could be punished for it. Walter Noll knew the whole story and didn't denounce him. That was how their real friendship began.

The origin of Walter Noll's conceptual approach in mathematics can be traced to as far back as the lessons of Rudolf Hohensee. The latter was an opponent of understanding this subject as a dull game of symbolic manipulations [1.17]. He taught Walter Noll that the language of science should be built with care. Once at home,

he filled a pair of glasses with two different fluids, one of which
adhered to the wall, the other didn't. He commented that there were
no "adherent fluids" in nature. The glass and fluid either adhere or
do not adhere to each other.

Among the documents related to Walter Noll's school time, an
important place belongs to his physics notebook. Its beginning can
be dated back to the spring of 1938, when Rudolf Hohensee began his lessons on physics [1.18]. On the first page of the notebook, we find a picture of the amusing demonstration of the sag-
ging wire. The teacher devoted the whole semester to mechanics.
Walter Noll learned from him about measurement of time, velocity, and force, hydraulics and gases, and mechanical, fluid, and gas
machines. He got to know the scientific biographies of the Krupp
family, Otto von Guericke (1602–1686), Blaise Pascal (1623–1662),
and Graf Ferdinand von Zeppelin (1838–1917), and some pages
from the history of aeronautics. In the winter semester of 1938/39,
Walter Noll had physics lessons on the rest of mechanics and also on
thermodynamics and acoustics. He studied the molecular hypothesis of matter, thermometry, phase transitions, elementary meteorol-
ogy, propagation of sound, and the structure of the human ear. In
the spring semester of 1939, Hohensee lectured on optics, which included theories of mirrors, lenses, and optical instruments, the propagation, reflection and refraction of light, and the structure of the
human eye. The grades for Walter Noll's homework evolved from
"good with minus" on May 27, 1938, to "very good" on March 25,
1939 [1.19].

In the tenth grade, Walter Noll became interested in chemistry
experiments. They fascinated him with their predictability. He remembered: "When I was about 15 years old, some of my schoolmates and I became interested in chemistry. We acquired all kinds of
equipment and chemicals, some of them dangerous, and we irritated
our mothers by attempting various experiments on the gas range in
the kitchen. Once I prepared a mixture of potassium chlorate and
red phosphorus in a small porcelain container. It exploded in my left
hand, and I still have two scars to remind me" [1.20]. In the winter
of 1939/40, the Second World War was already at the door. Most of

the young German teachers were called up for military service, and the quality of teaching deteriorated considerably [1.21].

Like many other pupils, Walter Noll saw his main problem after school as being "to save his head". When the war veterans came to the school to tell about their heroic deeds, Walter Noll felt not enthusiasm but fear [1.22]. Franz Noll and his family, when they were alone, often listened to the radio from London, although they could have paid with their lives for it. Franz Noll was sure that the war was lost for the Germans. The news of the victories of the enemies of Germany shocked and frightened Walter Noll. He was afraid of dying in the war. Walter Noll remembered: "In fact, the thought that I might not survive the impending disaster was never far from my mind" [1.23].

At his leisure, Noll learned to enjoy classical music and modern dance music: tango, waltz, foxtrot, and jazz. It wasn't difficult for him to get and to read books by Jews and German emigrants, because they were kept by many despite official prohibition [1.24]. Noll remembered: "I read a lot, not only popular science, literature and novels, but also philosophy such as Kant's 'Critique of Pure Reason'. I also participated in activities with my friends and classmates. I particularly remember a very enjoyable vacation trip that five of us made in the Harz mountains in the summer of 1942. I obtained my 'Abitur' [high-school diploma] on March 23, 1943, with the designation 'Gut' [cum laude]. At that time less than 10% of the people of my age obtained the 'Abitur'" [1.25].

In June 1943, Noll was drafted into the National Labor Service ("Reichsarbeitsdienst"). This organization was a state institution designed to overcome internal political and social conflicts in German society and to unite the German people under the powerful nationalistic idea of work for the motherland. The Reichsarbeitsdienst was aimed primarily at young Germans who had just finished their school education. The young men were to move earth, and the young women were used for agricultural work and gardening. The organization was as unpolitical as it could be in a totalitarian state. For example, members of the Reichsarbeitsdienst, who were simultaneously members of the Nazi party (NSDAP), were not allowed to act there in the service of the party and its subdivisions. According to the

German law of June 26, 1935, the Nazi leadership could not inter-
fere in the internal affairs of the Reichsarbeitsdienst. The connection
between the Nazis and the organization was put into the hands of its
commander Konstantin Hierl, a famous German military historian
and the creator of the basic ideas of the Reichsarbeitsdienst [1.26].

When Noll was drafted into the Reichsarbeitsdienst, its members
were used as building troops in France and Norway for the purposes
of the German Air Force and Army. The period of service in the
Reichsarbeitsdienst was from three to four months. Young Germans
were to receive a more paramilitary education there (without arms)
than they did in peace-time professional training. Noll remembered:
"We were shipped to Southern France, where I injured myself
slightly on the shin bone while unloading corrugated sheet metal.
The paramedics in my unit were totally incompetent, and the injury
led to a severe infection which almost cost me my right leg. In the
nick of time I was driven to the military hospital in Perpignan, where
a surgeon operated on me only a few minutes after my arrival. Later
I was shipped to a hospital in Alsace, where I stayed for several
months. After I recovered I was sent home just before Christmas
1943" [1.27].

On his dismissal from the National Labor Service in January of
1944, Noll gave a false home address, to which his call-up to the
German Army was sent. It was not until June 1944 that officials
discovered the error. He used this half a year to study mathematics
at the University of Berlin. The lectures were open to everyone,
and he did not need to enroll as a student. He took two lecture
courses and practical exercises: (i) "Algebra I" by H.L. Schmid,
which included a large portion of the contents of the book "Modern
Algebra" by B.L. van der Waerden; (ii) "Differential Equations" by
E. Schmidt [1.28]. Noll was working hard on mathematical books at
home. For example, he studied thoroughly the textbook "Theory and
Applications of Infinite Series" by K. Knopp [1.29].

On June 5, 1944, he sat before a select commission of the German
Army [1.30]. Due to his bad physical condition he was drafted into
the Air Force Signal Corps ("Luftnachrichtentruppe"). This corps
of German troops was created in 1933. Their main duties were air
traffic control, aircraft warning, the guidance of interceptors, and

radio reconnaissance [1.31]. Noll was sent to Augsburg, where a command of recruits was in training. He received special military training in the town of Auxonne near Dijon, France, and then in Weimar, Germany. In December 1944, he contracted diphtheria and had to stay for more than a month in an isolation ward. In February 1945, he was sent to the Russian front. The military situation there was not in favor of the Wehrmacht: Germans were slowly retreating before the overhelming majority of the Soviet Army. At the end of January, the latter reached the River Oder and stopped there due to strong German defence facilities and bitter resistance by the German troops. Noll arrived at the positions near Stettin, where detachments of the German Third Tank Army opposed the Second Belorussian Front under Marshal Konstantin K. Rokossovski. The heavy battles in East Prussia and Pomerania drew in the majority of Russian elite troops and reserves. For this reason, the Russian offensive at Stettin only began on April 16. After four days of bloody battles, the Soviet troops managed to cross the Oder. Noll destroyed his radio equipment and joined a group of retreating German soldiers moving to the West. Unlike Rudolf Hohensee, he managed not to become a prisoner of war of the Russians and to arrive at Lübeck. On May 1, 1945, he was captured there by British soldiers [1.32].

His British confinement continued only until the beginning of July. Noll knew that he had an aunt in Varel-Obenstrohe near Oldenburg. He received permission from the British occupation authorities to join her. In order to earn his living, Noll worked on the plots of his aunt and her neighbors, where they tried to grow vegetables. In December 1945, he was assigned to work at the Motor Plant in Varel [1.33]. Since he had no qualifications, Noll could only become an assistant worker. Some time later, he was promoted to clerk. Noll had to keep this employment until his return to Berlin. He remembered: "I devoted a large portion of my free time to systematic self-study of English, because I felt even then that one cannot get anywhere in this world without knowing English. I borrowed a textbook from a neighbor and listened to English lessons given regularly on the radio. By February of 1946 my knowledge of English was solid enough so that I never needed any systematic study again" [1.34]. At that time, he was unable to study mathematics as there were no mathematical books available [1.35]. Noll was eager to matriculate as a student at the Humboldt University in Berlin. He had wanted to begin his mathematical studies there in the winter semester of 1945/46. However, for well-known reasons, the teaching of mathematics could not begin there until the end of 1946. Noll's parents, who reestablished contacts with him in October 1945, advised him not to lose time. Then he wrote to his mother and asked her to bring his application documents to the Technical University of Berlin, which was to open its doors again in April 1946.

The situation in Berlin, where Walter Noll returned on March 12, 1946, and in Germany as a whole at that time, has been perfectly described in a book by Professor H.B. Wells, who followed Professor R.T. Alexander, the leading American reeducationist of Germany, in the post of Chief Educational and Cultural Advisor to the U.S. Military Government in Germany: "When I arrived in Berlin in the late fall of 1947, the city was still in ruins, I was told by persons who had been present three years earlier, at the time the German forces surrendered, that every aspect of the German society had completely collapsed. There was no responsible government – national, state, or local. . . . Industry was at a standstill, agriculture was disorganized, and trade and commerce were non-existent. Schools were closed and children roamed the streets. Looters robbed bombed-out homes and stores. The people were sullen, disappointed, dispirited. Germany was truly a beaten nation. All the important cities were from 50 to 75 percent destroyed" [1.36].

The reeducation politics greatly damaged the teaching of mathematics at the Technical University of Berlin. The university control officer fired many German instructors as "Nazis". The rest of the teaching staff was split and engaged in mutual intrigues. The head of the Department of Mathematics, Ernst Mohr, was a controversial figure. For some time, he was a fellow traveller of Nazi "German mathematics". However, he was arrested by the German secret police in 1944 as an opponent of the Nazi regime. Noll wrote about the university climate at that time: "Unfortunately, even at the Technical University, there are full professors who would willingly send the rest of the Jews into the gas chambers. ... The personal antagonisms in the Mathematical Department of the TU, especially the ... mean conduct of Prof. Haak, Herr Nitsche, and Prof. Mohr towards one another, give me further cause for doubt" [1.37].

Following enrollment, Noll rented a very small room in Berlin-Charlottenburg, which was situated, however, not far from the Technical University [1.38]. As a condition for matriculation, he had to work for 100 hours removing the ruins of the war [1.39]. He began his university studies in the summer semester of 1946. He took mathematics as his main subject and mechanics and physics as secondary subjects. He was lucky that Liese Hohensee, wife of his school teacher and friend, still worked at the Technical University. Noll remembered: "One of the instructors at the Technical University was Istvan Szabó, and I was introduced to him by Liese Hohensee, who at the time was employed as an assistant at the Technical University. Szabó recognized that I was a very good student and he arranged for me to be employed as 'studentische Hilfskraft' (teaching assistant) from the second semester on, which helped me financially during my years as a student in Berlin" [1.40]. In the first years, Noll found it difficult to study since he was starving all the time and freezing in winter, due to shortages of food and coal supplies to Berlin. The progress of Noll's studies was considerably dependent on the library of the Department of Mathematics. He remembered: "I probably learned as much, if not more, mathematics by the systematic reading of textbooks than by attending courses" [1.41]. He soon discovered that there was no great difference between the mathematical teaching programs at the Technical University and the Humboldt University [1.42].

In the summer of 1947, Noll took a course on social and political subjects, organized by the French Military Government. It was aimed at bringing together German students with other students from Europe. It lasted two weeks and was held in Freiburg, Germany [1.43]. In the winter of 1947/48, Noll and some other German students were allowed to take part in a discussion group. It was one of several groups organized on the request of Professor H.B. Wells, who wanted in this way to study the problems of the German education system. Noll's group met once or twice a month in the afternoon or evening. The majority of participants were war veterans, like Noll. Sometimes scholars and other German intellectuals were also invited. In order to create a relaxed atmosphere, these meetings were held at the houses of officials of the U.S. Military Government. Noll's group met at the residence of Professor H.B. Wells, "a comfortable four-storey, twenty-four-room house in the suburbs at 24 Am Kleinen Wannsee", the villa of a German movie magnate [1.44]. This luxury, together with stories about the American paradise across the ocean, had a great effect on Walter Noll. The meetings usually ended with a light snack. Noll remembered: "We were invited for sandwiches, coffee, and cake. It was difficult for us, starving students, to behave in a civilized manner and not to fight over the food like a pack of wolves" [1.45]. In the focus of interest at the meetings, there were the problems of the German population, the political future of divided Germany and Berlin, and the German education system. Peter Fraenkel, an aide to H.B. Wells and the informal US chairman of this discussion group, remembers that he also tried "to satisfy the immense curiosity of the German students about all subjects American ... and in particular about American education in general and American universities: organization, research, teaching methods and university student life in the United States" [1.46].

The preliminary exams, which took place in January 1948, were a simple exercise for Noll, who got very good grades. After that he began to spend much more time at mathematics lectures at Humboldt University [1.47]. In the summer of 1948, Noll participated in a two-week course by the French Military Government, which was held in Inzigkofen near Sigmaringen, Germany. Its topic was "The Peace". Directly after the course, Noll went to Great Britain with a group of German students to help farmers gather potatoes. They lived for two months in a labour camp in Yorkshire. The weekends were free, and the German students were able to enjoy sightseeing tours. During his stay in Great Britain, Noll visited London, Glasgow, and Edinburgh. At the end of his stay he spent two weeks on invitation at a family of German refugees in Cardiff, South Wales.

In 1949, Noll made use of his participation in the reeducation program organized by the French Military Goverment. He was granted a scholarship to study for one year at the University of

Paris. It was a considerable sum of money which was to cover the expenses for board, lodging and instruction [1.48]. He went to Paris before the beginning of the winter semester of 1949/50. The University offered him a small room south of Montparnasse, typical for students in Paris. In the atmosphere of postwar Germany, Noll had learned to be economical. He found a student canteen where the prices were low for regular students. To clear his head, he drank a lot of coffee and smoked many cigarettes. Outdoors, he always carried a large, old, leather briefcase, where he usually had a map of Paris and numerous dictionaries. In Paris, Noll had a small circle of French friends who introduced him to the cultural life of their capital. He got to know the architecture, paintings, and sculptures at Versailles, St. Denis, St. Germain, Fontaineblau, and Chartres. Together with other foreign students, he visited museums, movies, and concerts, where he often played the role of interpretor due to his very good knowledge of French. During the Easter vacation of 1950, Noll made a trip to Great Britain, where he stayed with the German refugee family who had offered him their hospitality in December 1948.

The studies at the Faculty of Science of Paris included lectures that were open to everyone, and little seminars for problem-solving. Since the number of students facing mathematics courses was large, the lectures were held in huge halls at the Henri Poincaré Institute (for example, in the "Salle Hermite"). The lecturers were the elite of French mathematics. For example, Professor G. Bouligand gave a course called "Differential and Integral Calculus". Walter Noll took only those courses which corresponded to the topics of the exams for the degree of Licencié ès Sciences. On the advice of a French student, Noll was also visiting the seminar on analytic functions of several variables at the École Normale Supérieure. It was there that he learned for the first time about the mathematics of N. Bourbaki. The seminar discussions usually took place among the French mathematicians H. Cartan, J.-P. Serre, and A. Borel. They were too technical to understand, but they left a deep impression on Walter Noll. Before his return to Germany, he bought all the available books of N. Bourbaki [1.49]. He was working hard at the mathematics library of the Henri Poincaré Institute. He wanted

to learn those parts of modern French mathematics that were little
known in Germany [1.50]. His first American friend, Ralph Raimi,
wrote that Noll's coordinate-free mathematics could be dated back
to his stay in Paris between 1949 and 1950. Raimi remembered: "I
had brought some books with me to Paris and had started reading
'Topological Groups' by Pontrjagin. Walter suggested that we read
it together. I can look back now and see why Pontrjagin's book
appeared to him so different from all that [applied mathematics].
Old-fashioned as it is, compared with what Bourbaki was publishing
at the time, it does have the abstract, 'minimalist' flavor Walter
thought had to be the essence of understanding. It had no coordinate
systems. A few years later, when Walter had come to America, his
ambition was already to characterize hydrodynamics axiomatically,
'without numbers'" [1.51]. Raimi also attributed the formation of
Noll's views on the role of mathematics in the physical sciences to
the time of his stay in Paris [1.52].

The friendship between Walter Noll and Ralph Raimi began in
Paris, where the latter arrived in August 1949 as a Fulbright stipen-
diary. Raimi didn't want to get a degree from the University of Paris
and his French was too poor to do so. However, he attended numerous lecture courses, three of them up to the end: (i) "Differential
and Integral Calculus" by G. Bouligand; (ii) "Integration" or "Measure" by J. Favard; (iii) "Linear Algebra and Analysis" by A. Lich-
nerowicz. Noll and Raimi were pushed together on the stairway of
a lecture hall. They were both strangers there and, following initial
conversations, they began to meet at lectures and to walk together in
Paris, talking English. Having learned that Noll was German, Raimi
told him of his Jewish roots and that many of his relatives in Poland
had been killed by Germans during the last war. Raimi remembered:
"This really took no further explaining, Walter taking it as a matter
of course that I held nothing against him on all this, and also as a
matter of course that despite the propaganda inflicted upon him in
his own youth he did not regard Jews as being different from other
people, and that I understood this. It may seem strange to have to explain such things today. 1949 was such a time, when Jews of America (and probably everywhere else) would not put their feet upon
German soil, and hated to hear the sound of the German language"

[1.53]. Noll's English wasn't good enough to discuss topics such as art or music, which were so interesting and attractive to Raimi, so they talked about politics. And they soon discovered that they were in good agreement on most of its problems [1.54]. On Sundays the two friends met often to discuss the book "Topological Groups" by L.S. Pontrjagin or an interesting mathematical problem. From his American friend, Noll could get fresh information about life and the prospects of a mathematical career in the USA, which strengthened his desire to go there [1.55].

On March 23, 1950, Noll was admitted to the exams for the degree Licencié ès Sciences at the University of Paris. In the last week of March, he passed the first exam "Algebra and Number Theory" with the grade "AB". In June, he took the last two exams: "Differential and Integral Calculus" and "Higher Geometry". He succeeded with the grades "B" and "AB", respectively [1.56]. The degree of Licencié ès Sciences in mathematics was officially granted to Noll about one year later [1.57]. After the exams, he made two sightseeing tours in France. The first one included the ancient castles in the Loire valley. The second trip went through a series of cities in Southern France: Avignon, Nice, Cannes, and Monte Carlo.

In August 1950, Noll returned from France and moved in with his parents, who lived at that time in West Berlin [1.58]. In the winter semester of 1950/51, he resumed his studies at the Faculty of General Engineering Sciences of the Technical University and his teaching assistantship, which had been interrupted by his stay in Paris. In order to earn more money, Noll managed to obtain such a teaching position not only at the Technical University, but also at the Free University. During the last two semesters, he obtained the remaining certificates to be admitted to the diploma examinations. His participation in the practical exercises during the whole period of study was evaluated as "very good". The oral diploma exams included the following subjects: group theory, higher technical mechanics, pure and applied mathematics, theoretical physics, and the theory of functions. Noll passed them successfully with good and very good grades [1.59]. His diploma was on the theory of functions. It was called "Reproducing Kernels, with Applications to the Theory of Analytic Functions" and achieved the grade "very good". On May

5, 1951, he was awarded the German degree of "Diplomingenieur" from the Department of Mathematics of the Technical University, which corresponded to the US degree of Master of Science in applied mathematics. In the introduction to his thesis, one can read the first reflections of Walter Noll's future concept of mathematics: "In the past hundred years mathematics went the way of more and more specialization and subdivision. However, recently there has emerged again a tendency for methodical unification. There has been a realization that only a few fundamental structures are at the basis of all branches of mathematics. The discovery of these structures leads to deeper insight and above all to considerable simplification. This development is necessary, for how else could the accumulating profusion of mathematical knowledge be preserved and transmitted" [1.60].

During his study at the Technical University, Noll made three good friends: Klaus André, Paul Winkler, and Dietrich Morgenstern.

Klaus André was born in Magdeburg, Germany, on June 2, 1929. He met Noll at the Technical University where he spent the summer semester of 1947. Together they visited the lectures by E. Mohr in mathematics and G. Hamel in mechanics. For the winter semester of 1947/48, André matriculated at the Humboldt University, which he later changed to the Free University in West Berlin. In 1953, he obtained a diploma in mathematics there. Noll had undoubtedly consulted André on his master's thesis, the subject of which was, as in Noll's diploma, reproductive kernels [1.61].

Paul Winkler obtained his diploma in mathematics at Philipps University, Marburg, in 1950. He moved to Berlin, where he became a scientific assistant at the Free University. He got to know Noll in the winter semester of 1950/51, when he conducted practical exercises and coordinated the work of teaching assistants. Noll was one of these teaching assistants [1.62].

In the summer semester of 1947, Noll became acquainted with a very talented mathematics student at the Technical University. His name was Dietrich Morgenstern. He was born on September 26, 1924, in Ratzeburg, Germany. In 1950 he obtained a mathematics diploma and began to work as a scientific assistant at the Technical University. Within two years he managed to get a Ph.D. there.

Noll and Morgenstern were very good friends. The latter had a motorcycle, and on it they sometimes made sightseeing tours out of Berlin [1.63].

2 Slave of Istvan Szabó

Walter Noll wanted only to be a mathematician. After receiving his diploma in May 1951, he was qualified to be a scientific assistant in mathematics. However, no vacancies were available at the Department of Mathematics at the Technical University. He was looking in vain for such a position elsewhere. Once he was offered such a post at the Free University, but this proposal was later withdrawn in favor of another candidate.

Noll's master's thesis was written under the scientific guidance of Ernst Mohr, the chief of the Department of Mathematics at the Technical University. As one of the diploma tasks, Noll was to consider a special case of the well-known Schwarz Inequality. He managed to find its solution only in the summer of 1951. Mohr suggested that they wrote a short article about it. This was ready in August 1951. Due to Mohr's authority, it appeared in the German journal "Mathematische Nachrichten" in March 1952 under the title "Eine Bemerkung zur Schwarzschen Ungleichheit". The paper deals with the following simple case of Schwarz Inequality:

$$\left\{ \int_a^b f(t)\mathrm{d}t \right\}^2 \le (b-a) \int_a^b \{f(t)\}^2 \mathrm{d}t \; ,$$

where $f(t)$ is n-times differentiable function on $[a, b]$.

Noll and Mohr derived the following formula:

$$\left\{ \int_a^b f(t)\mathrm{d}t \right\}^2 = (b-a) \sum_{v=0}^{n+1} C_v \int_a^b \{f^{(v)}(t)\}^2 [(b-t)(t-a)]^v \mathrm{d}t + (-1)^n R_n \; ,$$

where $C_v = (-1)^v [v! (v+1)!]^{-1}$;

$$R_n = \int\int_{a \leq x_n \leq \ldots x_1 \leq y_1 \ldots \leq y_n \leq b} \cdots \int \left\{ \int_{x_1}^{y_1} f^{(n)}(t) dt \right\}^2 dx_1 \, dy_1 \ldots dx_n \, dy_n$$

or

$$R_n = \int\int_{a \leq t \leq s \leq b} f^{(n)}(s) f^{(n)}(t) [(b-s)(t-a)]^n ds \, dt \ .$$

The mathematicians also posed two problems for the future, which they intended to solve:

(i) under what limitations on $f(t)$ do the numbers

$$\left\{ [v! (v+1)!]^{-1} \int_a^b \{f^{(v)}(t)\}^2 [(b-t)(t-a)]^v dt \ , \right.$$

$$\left. v = 0, 1, 2, \ldots \right\}$$

build a monotonously decreasing sequence;

(ii) find a class of functions $f(t)$ such that $R_n = 0$.

For the second problem, Walter Noll and Ernst Mohr proved that $R_n = 0$ for those functions $f(t)$ whose Taylor series, at every point of the closed interval $[a, b]$, had radius of convergence greater than $(b-a)/2$. The main result of this publication was included by E.F. Beckenbach and R. Bellman in their handbook on inequalities, published for the first time in 1961 [2.2].

Istvan Szabó, who was by 1951 the chief of the Department of Mechanics at the Technical University, offered Noll a scientific assistantship with him. Noll signed a two-year contract, which solved his financial problems. With the new salary, he could afford to rent a nice, large, furnished room in Berlin-Wilmersdorf [2.3]. He remembered that it was because of Szabó that he became involved with mechanics [2.4]. In October 1951, Noll began his first year as Szabó's scientific assistant. He had to teach engineering mechanics instead of mathematics, and did so without enthusiasm, considering it to be a displeasing task [2.5]. Very soon, Istvan

Szabó was informed of this by someone. The chief decided to use Noll for another type of work. In his position at the Technical University, Istvan Szabó was supposed to write mathematical and mechanical textbooks for engineering students. He usually wrote a number of pages at the weekend and, on Monday, he would give them to Noll. The latter was to look through the material, propose improvements and add examples and exercises. In reality, Noll had to write considerable parts of Szabó's books "Mathematische Formeln und Tafeln" and "Integration und Reihenentwicklungen im Komplexen, Gewöhnliche und Partielle Differentialgleichungen" [2.6]. As compensation, Szabó proposed Noll as a candidate to obtain a Ph.D. in the USA. This story will be told in the next chapter.

In February 1952, Walter Noll met Helga Schönberg, his future wife, at the student costume evening, "Roter-Zinnober Ball", in Berlin. It was love at first sight. They spent much of their free time together. Helga was German and came from a family of teachers. Her father was killed in the Second World War and her mother was a school teacher of English. On the eve of his first trip to the USA in September 1953, Noll and his girlfriend went together on a camping trip to France.

After his return from Bloomington in October 1954, Noll resumed his work for Istvan Szabó. He contributed largely to Szabó's textbook "Technische Mechanik" [2.7]. Noll's contract at the Technical University was to come to an end in a year. He tried to find a new position, but no vacancies were available for him in Berlin. Possessing great ambitions, the mathematician drew the false conclusion that he wasn't needed in Germany. It seemed to be a good alternative for him to look for a job in the USA, where he could expect support from a number of prominent mathematicians and mechanicians: C.A. Truesdell, D. Gilbarg, and R. Rivlin. Noll remembered that he had declined two offers of a university position in the USA in 1954. His father, who still had a great influence on him, advised him to leave Europe, where the growing hunger for power of the Soviet Union and the USA could soon ignite a new world war. He began to think seriously about emigration to the United States [2.8]. In December 1954, Noll had been offered a position of instructor of mathematics at the University of Southern California.

Then Noll wrote to Truesdell and asked him for advice. Truesdell offered him to send "blind applications" for a post-doctoral position at Harvard and Princeton Universities. Noll did so but unsuccessfully. This failure had severely disappointed him. However, Helga Schönberg soon managed to encourage her nervous, unbalanced, absent-minded boyfriend. In March 1955, Walter Noll accepted the offer from the University of Southern California.

On April 1, Helga Schönberg and Walter Noll got married in Berlin. They spent their honeymoon in Italy on a bus tour of Florence, Rome, Ravenna and Venice. After the marriage, Noll left his parents and moved into the flat of his wife and mother-in-law. He thought that in this way he could finally escape the influence of his father.

As Noll left for the USA in September 1955, he explained the move to Istvan Szabó in terms of the lack of opportunities in Germany. We can assume, knowing the character of Noll, that, on leaving his motherland, he promised Szabó to return if a suitable position could be found for him in Berlin. Szabó gave Noll an excellent reference where he warmly appreciated Noll's mathematical and pedagogic talents [2.9].

3 Truesdell

In 1952, Professor Georg Hamel, a famous German mechanician and a pupil of David Hilbert, received a letter from Professor Clifford Ambrose Truesdell, managing director of the Journal for Rational Mechanics and Analysis of the Indiana University in Bloomington, USA. Hamel was a member of its Editorial Board. In his communication, Truesdell asked Hamel to propose good young German mathematicians to study for a Ph.D. at the Graduate Institute for Applied Mathematics at the university [3.1]. Truesdell was eager to create the first scientific school of European rational mechanics in the United States. American students at that time were useless for his plans, so he decided to get pupils from Germany. It seemed very easy, since, on the one hand, Germany had an old tradition of rational mechanics and, on the other hand, German science was suffering severe financial problems which could force young German mathematicians to decide in favor of his offer [3.2]. Being a patriot of Germany and its mathematics, Hamel understood very well Truesdell's intentions, but he wanted to play along. He invited Istvan Szabó to discuss the matter with him. Hamel's ingenious plan was to use Truesdell's offer for the benefit of German science. They decided to recommend Walter Noll and Dietrich Morgenstern as suitable candidates, as they seemed to be patriotically minded and would get the American degrees and return to Germany thereafter. Szabó spoke to Noll, the first candidate. He gave him a pamphlet which explained how to arrange one's stay at Indiana University, and promised him a one-year leave from his duties at the Technical University. Szabó managed to conceal his real motives. He told his pupil that he couldn't offer suitable guidance for a doctoral dissertation in applied mathematics himself. It was done so cleverly that Noll believed him [3.3]. Truesdell received Walter Noll's application and accepted it. As an influential member of the staff at the Graduate Institute, he arranged for Noll to receive a research assistantship there [3.4]. He also advised him to apply for a Fulbright travel grant to cover travel costs between Germany and the USA.

At the beginning of September 1953, Noll arrived by sea in New York. He then proceeded to Rochester, where he stayed for several days at the house of his Paris friends Ralph and Sonja Raimi. From there, he took a train to Indianapolis and finally reached Bloomington by bus. Upon arrival in Bloomington, Noll got a place at the Rogers Center, where he shared a room with an emigrant from the Soviet Union. They could hardly communicate with each other, because of his roommate's poor English. In order to relax, Noll took part in the chess tournaments at the Rogers Center. He won one of these competitions, organized by Robert Byrne, a US grandmaster in chess, but Byrne then beat Noll in about twenty moves in a game which Byrne began without the Queen.

For his doctorate, Noll had to pass an exam in mathematics, prove his knowledge of German and French and defend a Ph.D. thesis. In October 1953, he successfully passed the oral doctoral examination in mathematics, mechanics and physics [3.5]. He persuaded the heads of the German and French departments of Indiana University to allow him to take his exams in these languages after the official deadline, which they did due to his late arrival at Bloomington. He did well in both exams. German was his native language, and he had improved his French during his long stay in Paris.

In December 1953, Noll took part in a sightseeing tour of Chicago, organized by the Indiana University. There he visited the Museum of Science and Industry, the offices of the famous Chicago Daily Tribune, a research laboratory of the Standard Oil Company, and the ABC radio and TV studios.

In the spring of 1954, he participated in a sightseeing tour of Washington, D.C., which included a specially arranged tour of the White House and the Capitol. As a German, he was once invited by a local church group to talk about the political situation in Berlin. At the Cosmopolitan Club of the Indiana University campus, Noll once unexpectedly met Peter Fraenkel, whom he knew from the discussion group in Berlin. Fraenkel was still an assistant to Professor H.B. Wells, president of the Indiana University. He arranged an invitation for Noll to a party at the president's home, which included a dinner, a concert by Arthur Rubinstein and a reception.

In the spring semester, Noll shared a room with Dietrich Morgenstern, who followed him to the Graduate School to obtain his second Ph.D. degree [3.6]. From the available graduate courses, Noll took partial differential equations, taught by D. Gilbarg, and hydrodynamics, lectured by Clifford Truesdell. He was the best student on the course of D. Gilbarg, and the latter even asked him to become his teaching assistant. Truesdell also conducted a seminar on statistical mechanics. He invited Noll to join it and to make a review of a paper there. The latter was happy to accept. He even managed to overdo the seminar task and to write an article on the connections between statistical and continuum mechanics. Truesdell was very much impressed and recommended it for publication in the Journal of Rational Mechanics and Analysis [3.7].

It is worth analysing this first contribution by Walter Noll to continuum mechanics. He believed that the laws of continuum mechanics could be deduced from the postulates of statistical mechanics. It was also a dogma for Noll that these postulates necessarily followed from the basic statements of quantum mechanics [3.8]. In 1954, he thought that a rigorous theory of continuum mechanics should be founded on statistical mechanics. He wanted to improve the mathematical rigour of the corpuscular model of matter in some contemporary chemical and physical theories. His particles were no longer molecules in the chemical sense, but material points. The influences of their rotations and degrees of freedom were excluded. Thus, there was no longer a difference between a mixture of material points and a chemical compound. Chemical and physical considerations had to play their roles only in the determination of potentials, which could be left to chemists and physicists. Noll wanted to obtain a purely mechanical problem. Instead of the common techniques of infinite series and delta functions, he gave closed integral forms for the concepts of his theory. Instead of the standard case, where a statistical system was influenced by external potential forces, Noll considered the forces to be dependent on the placement, time and velocity of material points. Walter Noll's way of proving the validity of the laws of continuum mechanics was very economical. He postulated the principle of preservation of probability in phase space. With its help, he introduced mathematical concepts for the following macroscopic state values: mass density, velocity, stress tensor, energy density, and heat flux density. He showed that the basic equations of continuum mechanics were then satisfied under certain "regularity conditions".

The most important problem of Noll's stay in Bloomington was to write a suitable Ph.D. thesis. He had to find an advisor at the Graduate Institute. As Istvan Szabó had promised to support his career at the Technical University after his Ph.D., Noll chose a dissertation on mechanics [3.9]. Clifford Truesdell was the most attractive figure among those staff members who worked actively in the field of mechanics at the Graduate Institute of Indiana University. Truesdell had arranged Noll's invitation, and he seemed to be convinced of Noll's mathematical talent. Noll could himself imagine the level of mathematical rigor in Truesdell's continuum mechanics. He asked Truesdell to be his Ph.D. advisor and the latter agreed [3.10]. Noll later stressed that the ideas of his thesis were entirely his own and that Truesdell had just given him a paper by S. Zaremba for guidance [3.11]. Noll's English was not yet perfect, and he spent much more time improving the syntax and style of his thesis than writing it. He could not avoid coordinate representations since the staff members of the Graduate Institute didn't know much about the mathematics of N. Bourbaki [3.12]. It took him about eight months to prepare the dissertation. From this fact, it is difficult to believe Truesdell who asserted that Noll knew nothing about its subject [3.13]. On July 26, 1954, he submitted a type-written copy of the Ph.D. thesis to the faculty [3.14].

We turn now to an analysis of Noll's dissertation "On the Continuity of the Solid and Fluid States" published in the Journal of Rational Mechanics and Analysis [3.15]. There are two competing contributions here. One is a short summary of the thesis, made by Noll himself and preserved in the announcement of his final examination on August 9, 1954 [3.16]. He pointed out two main achievements of his Ph.D. work: (i) he generalized the constitutive equations of

J.C. Maxwell and S. Zaremba for the amphimorphic materials; (ii) he introduced the principle of isotropy of space. Another contribution, belonging to Clifford Truesdell, was enclosed in Truesdell's letter to Georg Hamel on August 13, 1954. Truesdell pointed out the following main results of Noll's thesis: (i) Noll "formulated a general principle, which he called the isotropy of space, to be satisfied by all constitutive equations in classical field theories; from this principle he ... derived easily a number of important restrictions on such equations; (ii) his thesis concerned a particular class, which he showed to include both classical finite elasticity and two recent general theories of fluids; it may be regarded as a rigorous (i.e., properly invariant) embodiment of Maxwell's theory of visco-elasticity; (iii) the last part of his thesis obtained exact solutions in a number of special cases; these were interesting in that they show yield-like phenomena emerging as proved theorems, not special assumptions, in a visco-elastic theory consistent with all mechanical requirements for deformations of any magnitude" [3.17]. Undoubtedly these are the earliest reviews of the content of Noll's dissertation available [3.18].

Walter Noll didn't have time to look through a large quantity of scientific literature on the subject. He escaped this with the following remark: "Since all this literature concerns only various linearizations and 'approximations' and in generality does not go beyond Zaremba's work, it will not be considered here" [3.19]. In this case, it was a mistake, since he missed an important paper of J.G. Oldroyd, which included, in particular, something like Noll's principle of isotropy of space.

Noll described his view of the basic concepts of continuum mechanics. For him, a continuum body consisted of geometrical points, which could not penetrate each other. These points were called "material points". The measure, called "mass", was absolutely continuous. It guaranteed the existence of "mass density". The equation, describing the motion of a body, was

$$x^i = f^i(X^j, t),$$

$i = 1, 2, 3$; $j = 1, 2, 3$; X^j was the position of a material point in space.

After definitions of the kinematic concepts, Noll went to the principles of mechanics. He defined the Euler-Cauchy stress vector $s = s(x, t, n)$ as "a force per unit area acting through a surface element with unit normal n at the point x at time t" ({2.2}, p. 16). From the principle of momentum, he derived the formula $s = S(x, t)n$, where $S = S(x, t)$ was a linear transformation called the "stress tensor". He defined the principle of conservation of mass: "The mass of any part ... of the continuum is preserved in time" (ibid., p. 16). He deduced from its integral form the density equation

$$\rho = \rho_0 (\det A)^{-1}$$

and the continuity equation

$$\frac{d\rho}{dt} + \rho \operatorname{div} v = 0 .$$

Noll postulated then the principle of conservation of linear momentum:

$$\operatorname{div} S + \rho g = \rho \frac{dv}{dt} .$$

Furthermore, he assumed S to be a symmetric transformation [3.20].

Noll introduced, for the first time in the world literature on continuum mechanics, the concept of a "constitutive equation", which determined "the particular ideal material which we wished to study" [3.21]. These constitutive equations were differential or integral equations between mass density, motion and stress (ibid., p. 17). He differentiated between *two types of constitutive equations*:

(i) "for a given motion $f(X, t)$, the constitutive equation determined the stress $S(x, t)$ completely (ibid., p. 23);

(ii) for a given motion $f(X, t)$, the constitutive equation determined the stress $S(x, t)$ only up to an arbitrary initial stress distribution $S(x, 0)$" (ibid., p. 24).

He formulated the following principle of isotropy of space: "For any point $y(t)$ and any rotation $R(t)$, the transformation $\Phi\{y(t), R(t)\}$ defined by

$$x \to x' = y(t) + R(t)[x - y(t)] ;$$
$$f(X, t) \to f'(X, t) = y(t) + R(t)[f(X, t) - y(t)] ;$$
$$S(x, t) \to S'(x', t) = R(t)S(x, t)R^{-1}(t) ;$$

t, X, $\rho(X, t)$ unchanged leaves the constitutive equation invariant" (ibid., p. 19). A little further on, Noll also introduced the principle without mathematical formulas (ibid., p. 20). With its help, he could define homogeneous and homogeneous isotropic materials in terms of spatial transformations. He posed the problem for the future to give "a rigorous axiomatic definition of the words 'material' and 'constitutive equation'" (ibid., p. 24). He solved it, in part, later in 1958.

In his thesis, Noll applied the general theory to classify hygrosteric materials and solved numerous standard problems of the mechanical behavior of such materials.

On August 9, 1954, he sat before an examination committee for the degree of Doctor of Philosophy at the Graduate Institute of Indiana University. It included the following professors: C.A. Truesdell, D. Gilbarg, W. Gustin, V. Hlavaty, E. Konopinski, R.W. Thompson, and G. Whaples. Truesdell introduced Noll and his career to the audience. His thesis "On the Continuity of the Solid and Fluid States" was discussed by the members of the committee. The discussion mainly concerned two points of the dissertation: (i) Noll's generalization of the constitutive equations of J.C. Maxwell and S. Zaremba, and (ii) his principle of isotropy of space and its consequences for continuum mechanics. The debate ended positively. On September 7, 1954, the Trustees of Indiana University conferred upon Noll the degree of Doctor of Philosophy.

4 Noll's Odyssey from Los Angeles to Pittsburgh

On September 15, 1955, Helga Noll set foot for the first time on the American continent. This was in Montreal, Canada, where she arrived with her husband on board the freighter "Prins Frederik Henrik". From Montreal they made a long bus trip to Los Angeles. On the way, they lodged for one night at the house of Clifford Truesdell in Bloomington. In Los Angeles, the Nolls rented a small furnished flat, found through the University of Southern California [4.1]. On their first day in it they had to do without electricity. They had to wait for about a month for the container with their possessions to arrive from Germany. The lodging soon proved to be very noisy and Walter Noll could hardly concentrate on his work at home in the morning hours. Having saved enough money, the Nolls moved to another flat which was modern and had new furniture [4.2]. Helga Noll, who didn't want to become a housewife, found a job as a salesperson [4.3].

As it turned out, Walter Noll could reach his university office on foot. Much later he bought a cheap car and began to drive there. His new duty was to teach a low-level evening course in mathematics. Noll's situation at the Department of Mathematics at the University of Southern California was improving with time. His colleagues were extremely friendly and helpful, although some of them had earlier voted against offering him the position due to his German origin [4.4].

In December, Walter Leighton, chairman of the Department of Mathematics at the Carnegie Institute of Technology, wrote to Walter Noll and offered him an appointment there. Noll was interested [4.5]. In January 1956, the second letter came from Leighton, who offered Noll an associate professorship with a five-year contract. When Noll took this letter to Ralph Phillips, the chairman of his department, the latter called this offer unreasonable. Noll was shocked. When he came home, he found a letter from Istvan Szabó who offered him a position of "Diätendozent" at the Technical University in Berlin. Noll immediately wrote his answer to Szabó and declined it. At the same time, he already decided to accept Leighton's offer.

In January, Noll finished a paper called "Verschiebungsfunktionen für elastische Schwingungsprobleme". He wrote in it: "In this paper, both the displacement potential of Galerkin-Westergaard and Love's function are generalized in such a manner that they can be used not only for static problems but also for vibration problems of linear elasticity. It turns out that Love's function exists exactly for those problems in which the curl of the displacement field is parallel to a fixed plane. In order to render the method more applicable, displacements and stresses are expressed in terms of Love's function for general cylindrical and spherical coordinates" [4.6].

On February 4, Noll finally accepted Leighton's offer. Just a little later, he received the second letter by Istvan Szabó who offered him a full professorship in mechanics at the Technical University. Szabó promised that, as a full professor, Noll would have three full-time assistants, a secretary, and a lot of power and administrative responsibility [4.7]. Unfortunately for German science, it was too late. If the position in Berlin had become free a little earlier, Noll would have returned to the motherland. So, quite by chance, Germany lost one of its best mathematicians forever.

In March, Noll took part in the 36th regular meeting of the American Mathematical Society at the Pomona College in Claremont, California, with a report on differentiation in vector spaces.

During his work in the University of Southern California, Walter Noll showed himself to be a very good teacher and a shrewd researcher. Another young mathematician, Robert Finn, who shared an office with him at the Department of Mathematics, became his close friend. This friendship between Noll and Finn led to their collaboration on a paper in the field of fluid dynamics. This was finished in December 1956 and was called "On the Uniqueness and Non-Existence of Stokes Flow". Some time after they submitted it to the Archive for Rational Mechanics and Analysis, Clifford

Truesdell, the chief editor, pointed out to them that one of the main problems of the paper – the non-existence of two-dimensional Stokes flow – had already been solved in 1953. However, another main problem – the uniqueness of a three-dimensional Stokes flow – was solved by Noll and Finn for the first time. Their model described a physically reasonable class of flows. Moreover, the method of solution invented by Noll and Finn was simpler than the competing methods and could be transferred without essential change from two- to three-dimensional problems [4.8].

5 "Walter Noll, Our Teacher"

The history of Carnegie-Mellon University began in 1900, when Andrew Carnegie founded technical schools in Pittsburgh to prepare specialists for his huge industrial enterprises there. In 1905, these schools were transformed into the Carnegie Institute of Technology, where engineering degrees were granted [5.1]. In 1913, brothers Andrew and Richard Mellon founded a private research institute in Pittsburgh, which was to "conduct comprehensive research in the fundamental and applied natural sciences and to cooperate with industry in sponsored programs of research". Using the contemporary terminology, the Mellon Institute did research work in polymer chemistry, solid-state physics, and biochemistry [5.2]. On July 1, 1967, the two institutions – the Carnegie Institute of Technology and the Mellon Institute – joined to form the Carnegie-Mellon University. Today, this university has about 7000 students, who are being instructed by more than 500 teachers. It includes the following institutions: the former Carnegie Institute of Technology and the former Mellon Institute, the College of Fine Arts, the College of Humanities and Social Sciences, the Graduate School of Industrial Administration, the Mellon College of Science, the School of Computer Science, the School of Urban and Public Affairs, the Hunt Institute for Botanic Documentation, the Software Engineering Institute, the Nuclear Research Center (Saxonburg, PA), the Radiation Chemistry Laboratory (Bushy Run, PA), the Computation Center for Research in Computer Languages, the Educational Center for Curriculum Development, the Transportation Research Institute, and the Drama Department [5.3]. From the foundation of the college, its staff put the main emphasis on the following branches of mathematics: rational continuum mechanics, operations research, numerical analysis, discrete mathematics, differential equations, and optimal control [5.4].

Before I continue with the story of the life of Walter Noll, it is necessary to stop and to make a few introductory remarks. The time he spent in Pittsburgh was very peaceful and almost lacking in extraordinary events. His everyday life cycle included his home, the university, and shopping. When Helga became very ill, Noll took care of her, did all the housework and looked after the children. For these reasons it is sensible to concentrate further attention on his publications, on his participation in the scientific conferences, symposia, and meetings, and on his activities as a lecturer and propagandist of mathematics.

In October 1957, Noll's paper "On the Rotation of an Incompressible Continuous Medium in Plane Motion" {2.6} was published in the Quarterly of Applied Mathematics. Under the conditions that the continuous medium was incompressible and the motion was plane, he showed that Taylor's theorem (a technical result of continuum mechanics) remained valid for any material whatsoever, even for an anisotropic solid ({2.6}, p. 3).

From December 26, 1957, to January 4, 1958, Noll was one of the speakers at the International Symposium on the Axiomatic Method with Special References to Geometry and Physics, which was held at the University of California in Berkeley. In a report entitled "The Foundations of Classical Mechanics in the Light of Recent Advances in Continuum Mechanics" he presented his *first axiomatic of rational continuum mechanics* (see Sect. 8.1.1). It was published in the symposium proceedings in 1959 {2.10} and was reprinted in a selection of Walter Noll's papers in 1974 {1.4}.

In November 1957, R. Finn, a former collaborator and colleague of Noll's at the University of Southern California, arranged a lecture by Noll at the California Institute of Technology in Pasadena, California. It took place on January 7, 1958, and its topic was "Modern Theory of Continuum Mechanics".

At the end of June 1958, Noll and R.J. Duffin, professor of mathematics at the Carnegie Institute of Technology, finished a small joint article "On Exterior Boundary Value Problems in Linear Elasticity" {2.8}. Without delay, C.A. Truesdell accepted it for publication in the Archive for Rational Mechanics and Analysis [5.5]. In July 1958, Noll finished his classic work "A Mathematical Theory of the Mechanical Behavior of Continuous Media" {2.9}, which appeared in the Archive for Rational Mechanics and Analysis in the same year.

On May 13, 1958, Professor E. Sternberg invited Noll to give a talk at the Applied Mathematics Colloquium of the Brown University in Providence. Some details of this trip are known. Noll arrived in Providence on November 13, 1958. At 3:15 p.m., he had his first informal contacts with the participants of the colloquium over a cup of coffee in the Faculty Club. His lecture, entitled "A General Theory of Constitutive Equations", began at 3:45. The audience included advanced graduate students, teaching staff and many other interested people. At 6 p.m., Noll was a guest at a dinner in his honor.

At the end of November 1958, he gave a brilliant lecture on his axiomatic of rational continuum mechanics at the Graduate Institute of Mathematics and Mechanics at the Indiana University in Bloomington [5.6].

In the summer of 1959, C.A. Truesdell wrote to Noll and invited him to give a talk at the Rheology Section of the National Bureau of Standards of the U.S. Department of Commerce in Washington. Truesdell was interested in presenting rational continuum mechanics to its staff. Noll agreed. The official invitation followed in a telephone conversation between Professor Robert S. Marvin, the section chief, and Noll on June 23. This lecture took place at the beginning of July and was a great success.

At the beginning of September 1959, Noll flew to Paris, France, to participate in the work of the 4th International Colloquium on the Axiomatic Method in Classical and New Mechanics. The expression "new mechanics", proposed by A.C. Chatelet, the chairman of the colloquium, included the following disciplines: relativistic mechanics, quantum mechanics, and wave mechanics, and their practical applications from modern physics to industry [5.7]. Noll was invited due to his participation in the Berkeley Symposium on the Axiomatic Method, held from December 26, 1957, to January 4, 1958 [5.8]. He presided over the afternoon session of

the colloquium on September 9, 1959. His own lecture, called "La
Mécanique Classique Basée sur un Axiome d'Objectivité", began at
3 p.m. on September 10, 1959. It was followed by a remarkable discussion of *Noll's second axiomatic of rational continuum mechanics*, in which Hans Hermes, J.-L. Destouches, and R. de Possel took
part [5.9]:

H. Hermes:	Is it possible to cut a body into two? There are two possibilities: (i) either the matter has been divided in the body from the beginning, or (ii) there exists a process to cut a body into two.
W. Noll:	There are difficulties of a technical nature to give a formalism, which includes a discontinuity of this sort.
H. Hermes:	The problem is resolved from the beginning, when the bodies are described.
J.-L. Destouches:	Why do you separate the inertial forces from the others?
W. Noll:	Because one cannot secure a privileged frame (fixed stars are not unique).
R. de Possel:	This axiomatization isn't realizable in practice, since it supposes that two material points are necessarily distinct, when in practice it is sometimes obligatory to suppose them to disappear.

These critical remarks had a deep effect on Noll. The words of
Hermes attracted his attention to the problem of *fit regions* (see
Chapter 15), and Possel's attack forced him *to drop the principle of
inpenetrability* in the foundations of rational continuum mechanics.
Noll's lecture was published in the colloquium proceedings in 1963
{2.22}, and there is a reprint of it in the selection of his works
published in 1974 {1.4}.

On November 20, 1959, Professor R.S. Rivlin wrote to Walter
Noll and invited him to come to Providence to discuss the consequences of Noll's axiomatics of rational continuum mechanics for
changes in irreversible thermodynamics, which had been proposed
by Rivlin and B.D. Coleman. They met for five days at the Brown
University in December.

In 1959, four joint articles by Noll and Coleman on *fluid mechanics and thermodynamics* of materials were published.

At the beginning of January 1960, Professor Barry Bernstein, the
chief of the Applied Mathematics Branch of the U.S. Naval Re-

search Laboratory, invited Noll to Bethlehem. Bernstein and his staff were interested in knowing Noll's ideas on fluid mechanics. They asked him to present a report at the Strength of Solids Seminar of the NRL. This took place on February 12, 1960, and was called "Newtonian and Non-Newtonian Fluids". Noll described how Newtonian fluids approximated general fluids for slow motion [5.10]. The next day he flew from Washington to Baltimore. At the airport he was met by J.L. Ericksen and R.A. Toupin. They spent the rest of the day together discussing recent developments in rational thermomechanics. The next morning, Noll gave a lecture on "Thermodynamic Principles in Finite Elasticity" at the Mechanics Colloquium of the Mechanical Engineering Department of the Johns Hopkins University.

In 1960, Noll became a full professor and published only one paper, wrote in collaboration with Coleman. However, it was a sensational paper, "An Approximation Theorem for Functionals with Applications in Continuum Mechanics", where *fading memory*, a new fundamental concept of rational continuum mechanics and thermodynamics, was introduced. According to the Institute of Scientific Information in Philadelphia, this article was cited in over 310 publications in the Archive for Rational Mechanics and Analysis and thus became a citation classic in 1986.

In September 1960, Noll received a letter from Ralph Raimi, his old friend and professor of mathematics at the University of Rochester. Raimi invited him to give a talk at the Colloquium of the Department of Mathematics of the College of Arts and Science there and didn't hide his intention to try to win Walter Noll for the University of Rochester during his stay. He wrote, in particular: "We are very interested – still – in applied mathematicians for our staff – people like you. For this reason we would like your advice if you should come to speak here" [5.11]. Noll accepted the invitation. This colloquium talk took place in November 1960. Its topic was "Functional Analysis and Rheology".

In 1961, an American dream of Walter Noll's came true: *he and Helga naturalized and became Americans.*

At the end of the spring semester of 1961, Noll found free time to take up several invitations to lecture in the USA. His first visit was to the Institute of Technology at the University of Minnesota, to which he had been invited in September 1960. His lecture was devoted to his research on non-Newtonian fluids and took place on May 11, 1961. In August, he received an invitation from his friend J. Auslander to give a talk at the RIAS in Baltimore. He accepted this invitation in the middle of September and put forward the following three possible topics for his lecture: (i) "Axiomatic Minkowskian Chronometry", (ii) "Non-Newtonian Fluids", and (iii) "Convex Hulls and Gibbs' Thermodynamics" [5.12]. This lecture took place in November.

Despite a considerable teaching load, Noll found time for research work. In 1961, three articles, written in collaboration with Coleman, were published. One of them was a review of the joint results of Noll and Coleman on the *mechanics of viscoelastic fluids* {2.16}. The other two were devoted to *linear and non-linear viscoelasticity* ({2.17}, p. 18).

At the beginning of February 1962, Noll and Coleman finished a paper on the *steady extension of incompressible simple fluids*, which was first published in July that year {2.19,32,32R}.

In April, Coleman made a trip to Israel to take part in the International Symposium on Second-Order Effects in Elasticity, Plasticity, and Fluid Dynamics in Haifa. His joint report with Noll was an exposition of their *theory of simple fluids with fading memory* {2.26}. This was published in the symposium proceedings two years later.

At the beginning of April, Noll received a telephone call from Professor D.C. Leigh inviting him to give a seminar at the Mechanical Engineering Department of Princeton University. Noll arrived in Princeton on the morning of May 9. The seminar began at 4 p.m. and the audience included members of the teaching staff and graduate students in mechanical, aeronautical, chemical, and civil engineering. Most of them were totally unfamiliar with Walter Noll's works on rational continuum mechanics. The lecture was announced as "Recent Developments in the Theory of Non-Newtonian Fluids" [5.13]. After the discussions, Noll was invited for a traditional dinner with several mathematicians from Princeton University, at which they could continue to discuss his results in a more relaxed atmosphere. This visit to Princeton University brought him due recognition in the conservative circles of American mechanics, but it didn't have a significant effect on their views [5.14].

At the end of July, Noll formulated a nice *theory of viscometric and curvilinear flows* in the paper "Motions with Constant Stretch History" {2.20}. He arranged its publication in the Archive for Rational Mechanics and Analysis in the same year.

In the fall semester of 1962, Walter and Helga Noll moved to Baltimore, where he became a visiting professor at the Johns Hopkins University.

On October 26, he gave a lecture at the U.S. Naval Research Laboratory in Washington. His audience included the staff of the Institute of Fluid Dynamics and Applied Mathematics [5.15]. In December 1962, Noll became a father. Helga gave birth to a daughter, who was named Virginia.

In December 1962, he had received an invitation, signed by Professor R.J. Duffin, to give a lecture at the Department of Mathematics of the Carnegie Institute of Technology. He accepted in principle, but put it off until January 1963. He couldn't leave Helga and the child alone in Baltimore. He gave the lecture on January 24. It was devoted to *differential equations of finite elasticity* and comprised the results obtained by Noll during the first part of his stay at the Johns Hopkins University.

At the beginning of February, Noll and Coleman finished *a thermodynamical theory of elastic materials with heat conduction and viscosity*. C.A. Truesdell accepted it for publication in the Archive for Rational Mechanics and Analysis for the same year {2.21}.

Noll continued to serve as a visiting professor at the Department of Mechanics of the Johns Hopkins University in the spring semester of 1963.

In April, Noll was one of the principal speakers at the First International Symposium on Pulsatile Blood Flow, held at the Presbyterian Hospital in Philadelphia. In his report he reviewed the *contemporary mechanical theories of flow in tubes*. In the discussion of it, Noll could not defend his theory of rational fluid mechanics against the critical attacks of the conservative American mechanicians. However, they acknowledged several advantages of Noll's conceptual approach: its generality, its roots in the language of modern mathematics, and its obvious clarity for non-mathematicians. In October, a reading course of Noll's lectures on tensor analysis was published at the Johns Hopkins University (see Chapter 8). On November 11 and 12, Noll took part in the second meeting of the Society for Natural Philosophy on fluid dynamics.

In the fall semester of 1963, he returned to his duties at the Carnegie Institute of Technology. In the middle of November, he received an invitation from Professor W.R. Sears, the director of the Center for Applied Mathematics at the College of Engineering of Cornell University in Ithaca, New York. Sears didn't hide the purpose of this invitation: he and Professor Paul Olum, the chairman of the Department of Mathematics, wanted to win the famous Walter Noll for their university. The center employed two dozen staff members from mathematics, various branches of engineering, and natural sciences. Sears mentioned the following names: Wolfowitz, Kiefer, Kesten, Fuchs, Booker, Morrison, Salpeter, Ludford, Block, Mitchell, and others. The center was responsible for teaching graduate students of applied mathematics. In the invitation letter, Noll was offered a regular appointment at the Department of Mathematics with a position at the center, where he was to do most of the teaching work. Noll went to Cornell University on December 3. In his lecture, he described the *use of contemporary differential geometry in rational continuum mechanics*. His listeners were fascinated with his talk, which caused a long, interesting, stimulating discussion. In the evening, he was invited to a concert by the famous Jewish cellist Rostropovich and a party in his honor at the house of Paul Olum. However, Noll saw no sense in changing his position for another one at Cornell University.

In 1964, the American Mathematical Society held its annual symposium in Providence. Its topic was the Applications of Nonlinear Partial Differential Equations in Mathematical Physics. Walter Noll gave a lecture there on *equations of finite elasticity*. It was an expository work and appeared as a publication in the symposium proceedings in 1965 {2.28}.

In October 1963, Noll was invited to give one of the 1964 Mechanical Engineering Graduate Seminars at the Department of Mechanical Engineering of the University of Delaware in Newark. The invitation letter was signed by Professor Albert B. Schultz. Noll accepted the invitation and proposed the following list of topics for the seminar talk: (i) "Theory of Continuous Dislocations"; (ii) "Viscometry of Non-Newtonian Fluids"; (iii) "The Second Law of Thermodynamics in Continuum Physics"; (iv) "One-Dimensional Non-Linear Elastic and Plastic Waves". Schultz preferred the last topic and fixed the date of Noll's lecture as April 10, 1964. The seminar began at 3:30 p.m. and lasted about an hour. Noll had to be introduced to the audience by Millard Beatty, since very few had ever seen him.

On April 24, Noll gave a talk on mathematical models for space-time at the Department of Mathematics of the State University of New York at Stony Brook [5.16].

On September 24, 1963, Professor W.T. Sanders wrote to Noll and asked him to give a colloquium talk of his choice at the Department of Mechanical Engineering at Columbia University in New York. The scope of scientific activities at the department was very wide, from solid and fluid mechanics to plasma physics, both theoretical and experimental. Noll accepted the invitation at once and proposed that Sanders choose a topic from the four he had designed for his seminar talk on April 10, 1964, at the University of Delaware. On April 14, 1964, Sanders sent Noll a letter with final details concerning his lecture. He promised to send copies of the announcement of this lecture to all persons of his choice. Noll asked him to inform R. Toupin at the T.J. Watson Research Center in Yorktown Heights and also the staff of the Courant Institute of Mathematical Sciences of the University of New York. Noll arrived in Columbia at about 11:30 a.m. on May 13, 1964. At 3:30 p.m., Sanders invited him for a cup of coffee at the department's quarters. Noll enjoyed there his first contacts with his future listeners. In half an hour, he gave them a lecture on the *second law of thermodynamics in rational continuum mechanics*. After the discussions, Sanders invited Noll for lunch, where the fascinated listeners proceeded to ask the Pittsburgh mathematician questions.

In August, Noll became a father again. This time it was a boy, who was given the name Peter.

In September, Noll received a telephone call from Professor Irvin Glassmann, who invited him to give the Baetjer lecture at the Department of Aerospace and Mechanical Sciences of Princeton University. It was a great honor, and Walter Noll agreed without hesitation. He sent two possible topics for the lecture to Glassmann: (i) "Space, Time and Objectivity"; (ii) "The Second Law of Thermodynamics in Continuum Physics". The idea to invite Noll came from Professors Oscar W. Dillon, Jr., and Donald C. Leigh, who were eager to learn about his latest work on thermomechanics. At the lecture, which took place on October 15, the audience was composed mostly of graduate students and the staff of the department. Most of them had no idea about Noll's conceptual mathematics. The lecture was interesting and stimulating. It excited a long discussion, so that Noll had to stay overnight in Princeton.

In October 1964, Noll received a call and then a letter from Professor Edward L. Reiss from the Courant Institute of Mathematical Sciences at the University of New York. Reiss invited him to give a lecture at the Applied Mathematics Seminar there. Noll accepted the honor. This lecture took place on November 19. The audience at the seminar included research fellows, advanced doctorate students, staff members, and other interested persons. Noll called his lecture "The Use of Mathematical Concepts in Modern Continuum Mechanics". He presented in it a survey of his works on rational continuum mechanics at the Carnegie Institute of Technology. The lecture was considered very interesting and stimulating, but it had a modest effect. Noll wasn't even offered a position at the Courant Institute.

The year 1964 brought him three publications. One of them, which appeared in February, made public his *axiomatic of special relativity* {2.24}. The second article, written in collaboration with Coleman, was an important contribution to *finite thermoelasticity* {2.25}.

On January 14, 1965, Noll gave a talk on *viscometric flows of simple fluids* at the Graduate Seminar of the Chemical Engineering Department at the Carnegie Institute of Technology. It was not very successful. After that time, Noll preferred to avoid scientific contacts and discussions with other departments of his institution.

In February, he made a trip to Newark to give one of the principal lectures at the Seminar on Foundations of Physics at the Department of Physics of the University of Delaware [5.17]. The invitation, signed by Professor Mario Bunge, the chairman of the seminar and a world-famous philosopher of science, reached Noll at the end of September 1964. Bunge advised him on the content of his lecture [5.18]. The lecturers of the seminar were a very select group. For example, C.A. Truesdell was invited to speak on the foundations of continuum mechanics. Bunge also mentioned other invited speakers and their domains: Peter Havas was to lecture on general relativity, Henry Margenau was responsible for quantum mechanics, E.T. Jaynes was to speak on statistical mechanics, and Ralph Schiller was to describe the relation of quantum and classical physics. The title of Walter Noll's lecture was "Space-Time Structures in Classical Mechanics" {2.30}. In it he presented some recent ideas on space-time in rational continuum mechanics, relativity and quantum physics [5.19]. There was an important, interesting, characteristic correspondence between Bunge and Noll about this lecture. The chairman of the seminar wanted Noll to speak about his rational continuum mechanics. However, the Pittsburgh mathematician tried to avoid it. He wrote to Bunge on October 12, 1964: "I thought that my proposal to entitle my lecture 'Space-Time Structures' should not be inconsistent with a content that deals mainly with the axiomatic foundations of classical mechanics. There is perhaps some ambiguity in the term 'classical mechanics'. As far as I am aware, the work by McKinsey, Suppes, and others deals mainly with the classical mechanics that is general enough to include continuum mechanics. It turns out that for this more inclusive 'classical mechanics' (which has been developed extensively in the last 10 years or so) the classical space-time structure of Euclid and Newton is not as appropriate as another 'neo-classical' space-time. It is this 'neo-classical' space-time that I have worked on recently and that I should like to speak about". He couldn't tell Bunge about a silent rule of the members of the Society for Natural Philosophy: to avoid any sort of rivalry with C.A. Truesdell.

At the end of November 1963, Noll received a letter from Professor Robert Plunkett of the University of Minnesota. He was asked to give talks at seminars and colloquia of the mechanics departments of several educational institutions in the Midwest USA. The project which made it possible was called the Midwest Mechanics Seminar. It was elaborated in 1958 by representatives of these departments at seven Midwest U.S. institutions with the purpose of inviting outstanding specialists in different branches of mechanics to give lectures at their seminars and colloquia. Plunkett was present at Noll's lecture at the University of Minnesota on May 11, 1961. He remembered Noll as a "really outstanding speaker". This compliment had its effect, and Noll accepted the invitation. He planned this two-week tour for the spring of 1965. He also prepared a list of topics and abstracts of the lectures, which is shown in Table 5.1 [5.20].

On February 19, 1964, Noll received a copy of the announcement of the 1964–65 Midwest Mechanics Seminar Series from Plunkett, together with the names and addresses of the contact people at each institution, where he was to give a lecture. The other lecturers mentioned in the announcement were A.M. Freudenthal (Columbia University), K. Klotter (Technische Hochschule Darmstadt, FRG) and W.R. Sears (Cornell University). Some time earlier, the honor to do this lecture tour was granted to J.E. Adkins, J.L. Ericksen, J.N. Goodier, C.A. Truesdell and some other prominent American scientists. The schedule of Noll's tour is shown in Table 5.2.

In June 1965 Walter Noll gave several lectures on the basic concepts of rational continuum mechanics at the International Mathematics Summer Center in Bressanone, Italy. During his stay in Italy he received a telephone call from Dietrich Morgenstern who invited him to do a report at the Colloquium on Mathematics and Mechanics at the Mathematical Institute of the University of Freiburg, Germany. Noll couldn't refuse an old friend and agreed. On July 2, 1965, he arrived at Freiburg by train and gave a talk on the topic "Mathematische Behandlung des zweiten Hauptsatzes der Thermodynamik in der Kontinuumstheorie".

The year 1965 was a very happy one for Walter Noll. The Springer-Verlag published his fundamental treatise on *non-linear field theories of mechanics* {1.2}, written in collaboration with C.A.

Table 5.1. Topics and abstracts of Walter Noll's lectures for his lecture tour of the Midwest, USA, April 27 – May 7, 1965

Title	Abstract
Viscometric Flows in Simple Fluids	The class of viscometric flows includes simple shearing flow, Couette flow, Poiseuille flow, cone and plate flow, etc. A survey of the theory of these flows, steady and unsteady, in incompressible simple ("Non-Newtonian") fluids is given. Relevant experimental data are mentioned.
Space, Time, and the Principle of Objectivity	The classical space-time of Euclid and Newton is not the most appropriate for modern continuum mechanics. It should be replaced by a "neo-classical" space-time, in which there are only "instantaneous spaces" and no absolute space. In this setting, a generalized "principle of objectivity" appears as a natural restriction on possible constitutive equations.
The Second Law of Thermo-dynamics in Continuum Physics	In continuum physics, the second law takes the form of a restriction on possible constitutive equations. It is shown in detail how the law implies the positivity of the viscosities and heat conduction coefficients in Navier-Stokes fluids. The application of the law to more general constitutive equations is discussed.
Materially Uniform but Inhomogeneous Elastic Bodies	In simple bodies homogeneity and material uniformity are distinct concepts. The nature of this distinction is explained. The deviation from homogeneity is related to what has been called "continuous distributions of dislocations". Finite elasticity for bodies with such dislocations is discussed.

Truesdell [5.21]. Also, seven classical papers of his were reprinted in the volumes 8/II and 8/III of the International Science Review Series at the Gordon & Breach in New York [5.22]. In August 1964, Noll succeeded in proving an important theorem on the maximality of the orthogonal group in the unimodular group. He couldn't enjoy the laurels of priority for this elegant theorem, but this proof closed one of the few holes in his rational continuum mechanics. This result was published in 1965 in the Archive for Rational Mechanics and Analysis (see {2.27}).

Table 5.2. Walter Noll's lecture tour in the Midwest USA between April 27 and May 7, 1965

Date	Place	Topic
19.04.65	College of Engineering, University of Michigan in Ann Arbor	Materially Uniform but Inhomogeneous Elastic Bodies
27.04.65	Michigan Sate University in East Lansing	Space, Time, and the Principle of Objectivity
28.04.65	University of Wisconsin in Madison	Materially Uniform but Inhomogeneous Elastic Bodies
30.04.65	University of Minnesota in Minneapolis	The Second Law of Thermodynamics in Continuum Physics
05.05.65	Research Institute, Illinois Institute of Technology in Chicago	Space, Time, and the Principle of Objectivity
06.05.65	University of Illinois in Urbana	The Second Law of Thermodynamics in Continuum Physics
07.05.65	School of Aeronautics, Astronautics and Engineering Science, Purdue University in Lafayette	Space, Time and the Principle of Objectivity

Following 1965 Noll's mathematics quickly found its way to different branches of German science. In principle, it was a good time to get the famous Walter Noll back to Germany. His American citizenship could present no obstacle to offering him a full professorship at one of the leading German universities. However, German mathematicians and public education officials did not care about this. The 1945–1949 policy of reeducation had not changed the tradition of "self-torment" in German science, mentioned by the 1905 Nobel Prize winner, Professor Philipp Lenard [5.23]. However, it seriously damaged another tradition which made the scientific potential of Germany so strong: the feeling of "work for the motherland". According to their limited university education, German scientists only tried to understand all about Noll's mathematics and then to profit from this knowledge for their unimportant, avaricious, private interests. For example, on May 5, 1966, Professor Peter Werner invited Noll to do seminar talks at the Mathematical Institute A of the Technische Hochschule Stuttgart and at the Technische Hochschule Karlsruhe. This invitation coincided with Noll's participation in the Workshop on the Foundations of Physics in June–July 1966 in Oberwolfach, Germany. The growing interest of Noll in Germany had not been noticed by anybody, although the signs of it were very obvious. For example, Noll proposed as possible topics of his lectures the whole spectrum of his brilliant mathematical results: (i) "Die Struktur der allgemeinen Mechanik der Kontinua"; (ii) "Der zweite Hauptsatz der Thermodynamik in der Theorie der Kontinua"; (iii) "Gruppentheorie und Mechanik der Kontinua"; (iv) "Die Differentialgleichungen der endlichen Elastizitätstheorie". His lecture in Stuttgart was devoted to his rational thermomechanics. Peter Werner chose it because, according to the level of knowledge in Germany, the intersection field of theoretical mechanics and thermodynamics was very narrow. On June 22, Noll and his family arrived in Berlin. While his family enjoyed the cultural life of the former German capital, Noll went to Oberwolfach for the workshop. From his talks with German colleagues, he realized that his chances of getting a respectable professorship in mathematics at Göttingen, Berlin, München and other main universities were very bad. He then came to the ambitious idea of creating his own scientific school in the USA and, for this purpose, of inviting young and successful German mathematicians to Carnegie-Mellon University. He could actually meet such a person in Oberwolfach: Heinz Dieter Dombrowski (see Chapter 7). On July 7, Noll gave his lecture at the Technische Hochschule Stuttgart. On July 9, he repeated it at the Technische Hochschule Karlsruhe. This lecture was devoted to the structure of general continuum mechanics and was a contribution to the First South-West German Mechanics Colloquium, a new regular forum of scientists in technical mechanics from Stuttgart, Karlsruhe, Darmstadt, Saarbrücken and Freiburg.

After his return to the United States, Walter Noll was honored by the National Science Foundation to give four lectures about his rational continuum mechanics at the National Science Foundation

Conference on Continuum Mechanics in Blecksburgh. These lectures took place on August 17 and 18. Noll had the largest audience among the speakers and was celebrated there as *one of the leading American applied mathematicians* [5.24]. These flattering American honors on the one side and the cold reception given to his mathematics in Germany on the other made Noll abandon the idea of returning to Germany forever. His life took the usual course. In October he took part in the meeting of the Society for Natural Philosophy on the axiomatization of mechanics in Chicago. He was also satisfied with his publication in this year, since the Springer-Verlag had issued in 1966 his joint monograph "Viscometric Flows of Non-Newtonian Fluids: Theory and Experiment" {1.3}, written in collaboration with B.D. Coleman and H. Markovitz [5.25].

In 1967, Walter Noll was very often away from Pittsburgh. He was even able to find time to participate in the IUTAM Symposium "Mechanics of Generalized Continua" in Freudenstadt, Germany, where he tried to interest the young German mathematicians with his elementary *theory of inhomogeneities in simple bodies* {2.32},{2.32R}. He was able to meet a request from Professor John J. Deely, the chairman of the Mathematics Colloquium of the Sandia Corporation at Sandia Base in Albuquerque, New Mexico, asking him to give two colloquium talks there. Noll spoke on two subjects: the space-time of mechanics and his theory of simple materials. The colloquium talks took place on February 16 and 17. On March 13, he was able to satisfy another request: Professor R.D. Snyder and his colleagues at the West Virginia University invited him to give a lecture and to discuss the foundations of continuum mechanics with them. On March 31, Noll was in New Orleans, where he took part in the meeting of the Society for Natural Philosophy. On April 3, he gave a lecture on the space-time of mechanics to the Department of Mathematics at the Simon Fraser University in Burnaby, British Columbia. A month later Noll spoke about the principles of rational continuum mechanics to the solid-state physicists at the Battelle Memorial Institute in Columbus, Ohio. In the fall semester of 1967, he was able to leave Pittsburgh for a lecture at the Department of Engineering Mechanics at the North Carolina State University in Releigh, North Carolina. This took place on November 16, and he devoted it to the concept of simple body. This time he had a very nice, friendly, but incompetent audience. They invited him to come for a year to the North Carolina State University as a visiting professor and also to consider the possibility of staying there permanently [5.26]. These trips and lectures took too much of his time and energy and he managed to write only one paper in 1967. This he finished during the summer vacation, when he was free from teaching, and it was devoted to the theory of dislocations (see {2.31}).

Walter Noll tried to combine his lectures with pleasure. Sometimes he accepted invitations for lectures purely because they were to take place in a new, interesting city with a well organized sightseeing program. In February he arrived in Florida where the weather was much better than in Pittsburgh. He gave a report on his theory of dislocations to the Department of Engineering, Science and Mechanics at the College of Engineering of the University of Florida in Gainesville. The audience included the staff of the Continuum Mechanics Group and other engineers, mathematicians and physicists. In May he came to Washington, where he again visited the Capitol, the art galleries and the museums. On May 6, he gave a colloquium talk at the College of Engineering in the Department of Aeronautics and Astronautics at the University of Washington. It was the same lecture as in Gainesville, and he needed little time to prepare for it. Noll also tried to find out whether he could get a good position at the college. On May 13, he received an answer from Professor E.H. Dill that they would be delighted to have him as a visiting professor, but not more. On May 7, Noll was received at the Boeing Scientific Research Laboratories. It was more of a social visit; it is difficult to believe that the key scientific personnel of the company was interested in listening to Noll's lecture "Time, Space, and Fast Interstellar Travel", written for freshmen and sophomores [5.27].

Noll got another chance to try to leave his *Pittsburgh captivity*. On May 17, he gave a lecture entitled "Mechanics Based on Neo-Classical Space-time" at the Applied Mathematics Colloquium of Princeton University [5.28]. This lecture was very carefully prepared and he included in it the whole arsenal of his progressive mathematical theories. In the lecture announcement he gave particular emphasis to space-time concepts. He wrote for example: "In order to clarify certain invariance principles occuring in the mechanics and thermodynamics of continuous media, it is useful to base these disciplines on neo-classical, rather than classical, space-time. The distinctions between classical, neo-classical, and relativistic space-time structures will be discussed. An outline of the mathematical theory of neo-classical mechanics will be given" [5.29]. However, he was deceived in his hopes. The lecture was pleasantly received by the audience but no offers of a position at Princeton University followed.

In 1968, Walter Noll made his first trip to a communist state. Professor H. Zorski invited him to give a lecture at the Instytut Podstawonych Problemow Techniki of the Polish Academy of Sciences in Warsaw and another lecture in Cracow. Walter Noll accepted although he had to pay travelling expenses from his own pocket. The first lecture took place in Warsaw on May 28. It was called "Mechanics and Thermodynamics Based on Neo-Classical Space-Time". On May 31, Walter Noll repeated it in Cracow. He spent the next month at the Technische Hochschule Karlsruhe in Germany. Professor W. Gunther arranged a visiting professorship for him there. On June 24, Walter Noll joined his wife and children in Italy. Before he could enjoy his vacation he had to give a lecture at the University of Pisa [5.30].

Among the documents relating to 1968, one can find proof that Noll remained devoted to his friends, despite the fact that they sometimes profited from his unwillingness to defend his priority [5.31]. The following story is characteristic of him. On February 15, 1968, K.C. Valanis asked Walter Noll to give one of the three principal lectures at the 11th Midwestern Mechanics Conference which was to be held at the Iowa State University in Ames on August 18–20. In his answer of February 26, Walter Noll modestly declined this invitation. Under a false pretext *he passed the proposed honor of a principal lecturer to a friend* [5.32].

In June 1968, Noll was invited to join the program of the Southwest Mechanics Lecture Series 1968. The group of lecturers also included A.J. Durelli, G. Herrmann, R.E. Kabala, S.J. Kline, R.T. Schield and C.E. Taylor [5.33]. The topic of Noll's lectures was "Where Does the Stress Tensor Come from?" The schedule of his lectures is given in Table 5.3. In Dallas Noll's lecture was for the first time in his life televised over a closed circuit TV-system in order "to make contact with all the major industries in this area as well as to get participation in the studio itself" [5.34].

Walter Noll was a very responsible mathematician and, in 1968, he spent the most part of his creative hours learning big portions of modern algebra for his collaboration with H.D. Dombrowski. His only annual paper, entitled "Quasi-Invertibility in a Staircase Diagram", appeared as a research report of the Department of Mathematics of Carnegie-Mellon University (see {2.33}). Fortunately, three old papers of his were reprinted in the book "Continuum Theories of Inhomogeneities in Simple Bodies", so that his annual report to the university could look better [5.35].

Table 5.3. Schedule of Walter Noll's lectures in Texas

Date	University	Place
October 8, 1968	Louisiana State University	Baton Rouge
October 9, 1968	University of Houston	Houston
October 10, 1968	University of Texas	Austin
October 11, 1968	Southern Methodist University	Dallas

Although there were no joint publications by Walter Noll and B.D. Coleman following 1966, they continued to discuss the topics of thermomechanics from time to time. These contacts became very intensive between November 1968 and February 1969. However, Noll was deeply occupied with his algebraic research. Like many other people, he was unable to work on several topics with the same intensity. The problem of a proper definition of the constitutive equation of thermomechanics was too difficult for Coleman, and even consultations with C.A. Truesdell and his colleagues at the Johns Hopkins University during Noll's stay in Baltimore in November 1968 did not sufficiently clarify the issue.

In March 1969, Noll received an invitation from Professor Edgar Enochs to give a colloquium to the Department of Mathematics at the College of Arts and Sciences of the University of Kentucky in Lexington. This talk took place on April 9, 1969. Its topic was algebraic, presumably, "Annihilators of Linear Differential Operators" [5.36].

During the spring semester of 1969, Walter Noll served as a visiting professor at the Israel Institute of Technology in Haifa.

On August 21, Professor Eli Sternberg, the chairman of the Solid Mechanics Seminar at the California Institute of Technology in Pasadena, invited Noll to give a talk there in January–February 1970. Since the travel expenses from Pittsburgh to Pasadena could be high, he suggested Noll also gave lectures at several other institutions in California (see Table 5.4).

Table 5.4. Schedule of Walter Noll's lectures in California, January 1969

Date	Institution	Topic
26.01.69	University of California in San Diego	Where Does the Stress Tensor Come from?
27.01.69	California Institute of Technology in Pasadena	Materially Uniform but Inhomogeneous Elastic Bodies
28.01.69	University of California in Los Angeles	On the Concept of a Simple Material
29.01.69	Stanford University	The Concept of Simple Material
30.01.69	University of California in Berkeley	Materially Uniform but Inhomogeneous Elastic Bodies

On January 26, 1970, Noll arrived at La Jolla, where he gave the lecture "Where Does the Stress Tensor Come from?" to the Department of Aerospace and Mechanical Engineering Sciences at the University of California. The audience included graduate students in engineering mathematics and mathematicians, interested in continuum mechanics. The talk began at 3 p.m. It was followed by a long discussion on various problems of elasticity, materials with memory and elastic stability between Noll and Nemat-Nasser.

On June 13, 1969, Professor Robert H. Owens, the chairman of the Visiting Lectureship Program Committee at the Society for Industrial and Applied Mathematics (SIAM), wrote to Walter Noll and asked him whether he would be interested in becoming a

SIAM-lecturer in 1969–1970. The program was aimed at sponsoring lectures of applied mathematicians at the American institutions without doctoral programs. Every lecturer was to prepare talks at all university levels: general, educational, freshman-sophomore, junior-senior and advanced. He was also to give at least six lectures during the time-period. Noll was interested to get this job which would bring him away from Pittsburgh and promised an additional purse.

In the SIAM information brochure on lectures in 1969–70, Walter Noll's name stood together with those of J.H. Ahlberg, W.F. Ances, D.J. Benney, G.R. Blackley, W.H. Bossert, W.S. Dorn, R.A. Gaggioli, R.T. Gregory, H. Halkin, J.H. Halton, G.H. Handelman, G. Leitmann, J.M. Ortega, R.H. Owens, S.V. Parter, R.J. Plemmons, F. Proschan, T.R. Rogge, R. Rosen, I. Stakgold, A. Strauss, B. Stubblefield, J.F. Traub, W.R. Utz and A. Wouk. The topics of Noll's SIAM lectures are shown in Table 5.5.

On September 3, 1969, Professor Richard C. DiPrima wrote to Noll asking him to send his mailing address to several nearby institutions of his choice which had no doctoral program in mathematics. After the information brochure had been distributed, Noll received many invitations, but he rejected those from institutions which were more than two hours by car from Pittsburgh. The crime rate in Pennsylvania rose to a new record and he was so afraid for his wife and children that he couldn't leave them alone even for one night.

Table 5.5. Topics of Walter Noll's SIAM lectures in 1969–1970

Level	Topic
General	The Several Roles of Mathematics in the Physical Sciences
Freshman-Sophomore	Relativistic Time and the Feasibility of Fast Interstellar Travel
Junior-Senior	Axiomatic Analysis of Classical, Neo-Classical and Relativistic Space-Times
Advanced	(i) Modern Mathematical Theories of the Behavior of Materials (ii) What is the Second Law of Thermodynamics? (iii) Tensor Analysis as a Branch of Pure Algebra (iv) The Structure of Linear Transformations

On December 18, he agreed to give a lecture at the Mathematics Department of the Geneva College in Beaver Falls. The lecture took place on January 12, 1970. It was devoted to relativity theory. Noll remembered his stay at Beaver Falls: "There were about twenty people in the audience. They were attentive and there were a few questions at the end. I am not quite sure about the real effect [of the lecture]. I had some discussions with the faculty during lunch and before my lecture. They asked me for advice on curriculum, choice of textbooks, and similar matters" [5.37].

On December 18, 1969, Noll agreed to give a lecture at the Mathematics Department of Shippensburg State College. It took place on February 16, 1970. The audience included undergraduate students of mathematics and computer science and members of the teaching staff. He gave, in fact, two talks. The first one, called "The Several Roles of Mathematics in the Physical Sciences", took place at 1:30 p.m. After dinner, he gave another lecture on relativistic time and the feasibility of fast interstellar travel. Noll became so tired that he was unable to drive back. He had to lodge for the night in a guest room on the college campus. He remembered his visit to Shippensburg: "I gave two lectures. The general lecture in the afternoon was poorly attended; there were only about eight people, all faculty members. The lecture in the evening had a much better attendence; about thirty people were there. There was much discussion after this lecture. Apart from the lecture, the most useful part of my visit was a bull-session I had with several faculty members and one very enthusiastic student. It went on until about 11:00 p.m." [5.38].

On December 2, 1969, Noll accepted an invitation to give a SIAM lecture at the Department of Mathematics at Millersville State College. His lecture, called "Relativistic Time and the Feasibility of Fast Interstellar Travel", took place on February 17, 1970. At the college, Walter Noll was honored as a mathematical celebrity. He was openly called a "prominent applied mathematician" by the staff and students. His audience was very large; about 100 people came to see him and to listen to his lecture. Contrary to his earlier experience, the discussion wasn't directed against his scientific views. On the contrary, *every word of his was taken as a gift from God*. Noll remembered: "This visit was perhaps the most effective so far. The faculty members were very relaxed and we had a lot of interesting discussions on such topics as 'What is Applied Mathematics?' and the role of applied mathematics in the curriculum. In the early afternoon I met with a group of about 10 senior class students who asked me a lot of questions" [5.39].

On October 24, 1969, Linda L. Clark invited Walter Noll to give a lecture at the Marquette University and at the University of Wisconsin-Milwaukee. He accepted. On April 6, 1970, he gave a lecture "What is the Second Law of Thermodynamics?" at the College of Engineering of Marquette University. He remembered: "The lecture was organized by the student body of the College of Applied Science and Engineering. About sixty people, mostly students, attended. There was great interest, and informal discussions with the students had to be cut short" [5.40]. On April 8, Noll gave a talk about his rational mechanics at the College of Applied Science and Engineering (CASE) at the University of Wisconsin-Milwaukee. The audience included staff members from the Mathematical, Physical, Mechanical and Materials Departments of the college. Before the principal lecture he agreed to do a lecture on relativity theory and space travel specially for the members of the CASE Student Organization of the university. Noll remembered his stay at Milwaukee: "The lecture was attended mostly by faculty [members] who were interested in modern continuum mechanics (about 20 people). At least one person came over from the University of Wisconsin in Madison. The discussions had to be cut short because I had to catch the plane home. ... I believe the visit to Milwaukee was very effective. I had close contacts with about five faculty members from Marquette and UWM [University of Wisconsin-Milwaukee]. We met for dinner on April 7 and lunch on April 8 and thus had ample opportunity for discussion, mostly about mechanics and thermodynamics and about the role of mathematics in these disciplines" [5.41].

On December 2, 1969, he accepted an invitation to give a SIAM lecture at the Department of Mathematics at Denison University in Granville, Ohio. This lecture took place on April 14, 1970, and it was about relativistic time and fast interstellar travel. The audience included seven staff members and some twenty senior and junior mathematical students. Noll remembered about his stay in Granville: "[The] audience ... seemed to follow the lecture attentively. There was no question and answer period at the end. I talked with several members and with students both before my lecture and during dinner afterwards. Unfortunately, there was very little discussion of either the substance of my lecture or of problems pertaining to the teaching of mathematics, curriculum matters, etc. This was too much of just a social visit" [5.42].

On May 6, Noll gave his final SIAM lecture to the Department of Mathematics at the Case Western Reserve University in Cleveland, Ohio. Its topic addressed again the freshman-sophomore level, although the audience included both staff members and undergraduate students.

On May 21, Professor W.E. Langlois wrote to Noll. He intended to organize a seminar series on the subject "Mathematics from the Sciences" at the Department of Mathematics of the Notre Dame University, Indiana. Langlois had attended Noll's lecture at the University of California in Berkeley on January 30, 1969, and he remembered Noll to be a lucid, entertaining, informative speaker. He asked him to consent in principle to giving a lecture at the proposed seminar series. Thus he could approach the potential sources of financing of the project. Walter Noll agreed, but finally, it came to nothing. Although Langlois did his best to find financial support for the lectures of the famous American mathematician Walter Noll, his efforts were in vain [5.43]. Walter Noll was very dissatisfied that his name didn't have as much weight as it should in the American university circles. On November 20, Noll wrote to Langlois and apologized that "nothing had come up that would bring him to the vicinity of South Bend". He mentioned the coming SIAM lecture tour of 1971, where he intended to take part in the spring semester, and gave an uncertain promise to visit the University of Notre Dame on the route of one of the lecture trips.

On May 29, 1970, Professor W.K. Tso invited Walter Noll to give a lecture on applied mechanics at the Department of Civil Engineering and Engineering Mechanics at McMaster University in Hamilton, Ontario. This lecture took place on October 28, 1970. It was called "On the Basic Concepts of Mechanics". The audience included staff members and graduate students in different branches of engineering science.

In 1970, two of Noll's articles were published. One of them, finished in February 1968 and called "Representations of Certain Isotropic Tensor Functions" {2.34}, appeared in the prestigious journal "Archiv der Mathematik" and was intended mostly for a German audience. At the end of February 1970, Noll completed an important paper on statistical mechanics, entitled "On Certain Convex Sets of Measures and on Phases of Reacting Mixtures" {2.35}, which was published without delay in the Archive for Rational Mechanics and Analysis.

On June 11, Richard C. DiPrima sent Walter Noll thanks for his contribution to the 1969–70 SIAM Visiting Lectureship Program. He invited Walter Noll to continue his service as a SIAM lecturer in the academic year 1970–71. The latter agreed. In November he sent DiPrima a list of nine institutions without doctoral programs, which might be interested to participate in the SIAM Program. The list of topics of his lectures remained the same (see Table 5.5) [5.44]. Together with Noll, the SIAM lecturers group of 1970–71 also included the following outstanding mathematicians: J.H. Ahlberg, W.F. Ames, D.J. Benney, Z.W. Birnbaum, W.H. Bossert, K.L. Deckert, B. Epstein, R.A. Gaggioli, J.M. Gary, R.T. Gregory, G.H. Handelman, W.L. Harris, SR., S.C. Lowell, Z.C. Motteler, J.M. Ortega, S.V. Parter, R.J. Plemmons, T.R. Rogge, D.A. Sanchez, G.R. Sell, I. Stakgold, A. Strauss, R.A. Struble, B. Stubblefield, R.M. Thrall, J.F. Traub, and W.R. Utz.

The first invitation for a SIAM lecture, dated January 22, 1971, came to Noll from Professor J. Tinsley Oden, chairman of the Division of Engineering at the University of Alabama in Huntsville. Walter Noll's lecture was called "Modern Mathematical Theories of the Behavior of Materials", and he gave it on February 11. The talk and active discussions with staff members lasted about two hours. After lunch he was invited to make a tour of campus facilities. He remembered this lecture: "The audience consisted of about twenty people, mostly faculty [members] and graduate students. They had a very good background and seemed to have no difficulty in following the lecture. I believe the visit was one of the most effective I ever made. I had many interesting discussions and actually gave an additional one-hour lecture on my recent research to a group of four people" [5.45].

On January 21, Noll received a telephone call from Josephine Story at the Department of Mathematics of Chipola Junior College in Marianna, Florida. She invited him to give a SIAM lecture there. Walter Noll agreed. On the morning of February 12, he arrived at the Chipola Junior College and used much time consulting staff members about their curriculum problems. After that he organized a discussion with them on the role of mathematics in the physical sciences. Noll's lecture was called "Relativistic Time and the Feasibility of Fast Interstellar Travel", and it attracted about eighty students and staff members. Noll was a little dissatisfied that there was no question and answer period after the lecture. However, after lunch he was able to have a nice chat with some interested mathematics majors on the professional opportunities in industrial and applied mathematics [5.46].

In his letter of February 19, Professor Richard A. Howland invited Noll to give a SIAM lecture at the Department of Mathematics and Astronomy at the Franklin and Marshall College in Lancaster. Walter Noll arrived in Lancaster on March 8. His freshman-sophomore lecture, called "Relativistic Time and the Feasibility of Fast Interstellar Travel", began at 4 p.m. in the Kauffman Lecture Hall on the college campus. The audience was again very large. Noll was very pleased with the warm reception and attention his personality and mathematical works received. He was especially pleased to give an interview about them to a local newspaper. In his report to SIAM, he called the college "a good school" [5.47].

On March 9, Noll gave two SIAM lectures at the Villanova University. The first lecture on relativistic time and space travel was given before lunch. There was no discussion after the lecture, but Walter Noll was invited for coffee and cakes, where he could answer the questions of the staff and students in a more relaxed atmosphere. The second lecture, called "Axiomatic Analysis of Classical, Neo-Classical and Relativistic Space-times", was held at 4:30 p.m. Noll's lectures at the Villanova University were admired by the listeners [5.48]. He remembered: "There was a large audience in the first lecture (about fifty) and a somewhat smaller audience (about twenty) in the more technical second lecture. The response was very good to both lectures. I believe the visit was quite effective. In addition to the lectures, I had private discussions with faculty [members] and students on various topics of mutual interest" [5.49].

In March Noll was asked to give three SIAM lectures at the Mathematics and Engineering Departments of the University of Arizona in Tucson. The first lecture took place on May 3 at the Joint Graduate Seminar of Aerospace and Mechanical Engineering, Civil Engineering and Engineering Mechanics Departments. The topic of the lecture was "Modern Mathematical Theories of the Behavior of Materials". At 4:40 p.m., Noll gave a colloquium talk on the flow of non-Newtonian fluids at the Mathematics Department. The following day he gave his third lecture to participants of the Aerospace and Mechanical Engineering Seminar. Its topic was "Relativistic Time and the Feasibility of Fast Interstellar Travel". He wrote about his lectures in Tucson: "Lecture one was attended by about thirty faculty [members] and students with interest in modern mechanics. Lecture two was more technical, with about ten people attending. Lecture three was attended by about hundred people, mostly undergraduates. This is the longest visit I have ever made. I came in contact with people from Engineering and Mathematics, and I discussed many topics with many people. I was consulted extensively by a doctorate candidate. The visit was meticulously organized. I believe it was extremely successful" [5.50].

On February 10, he agreed to give a lecture at the Mechanical Engineering Department of the University of Colorado in Boulder. This took place on May 6 and was devoted to Noll's mathematical theories of the behavior of materials. The level of the lecture was too high for the audience, and very few could understand and discuss it afterwards. Noll remembered that he had discussions with about three or four faculty members who asked him questions and elicited his opinions. They also discussed their own research work with him [5.51].

On February 10, Noll decided to give the last SIAM lecture to the Department of Mathematics at the University of Toledo, Ohio. His lecture, called "What is the Second Law of Thermodynamics?", took

place on May 12. It was preceeded by an introductory word, given
by H. Westcott Vayo, in which he made flattering remarks about the
speaker. However, Noll didn't notice any students in his audience;
they were all faculty members. There was a lazy discussion after the
lecture which concerned mostly the applications of mathematics in
physics and engineering. Mostly pure mathematicians, they had a
very limited understanding of the problems of rational mechanics.
Noll did find a staff member who was genuinely interested in
applications to physics, and talked with him at length [5.52].

On September 22, Professor William W. Adams invited Walter
Noll to give a colloquium talk to the Department of Mathematics of
the College of Arts and Sciences at the University of Maryland in
College Park. This took place on October 22 and was entitled "A
Mathematical Theory of Physical Systems with Memory". This was
a completely new theory, and Noll was able to discuss it there with
the best American specialists on the subject, including his friend
Professor J. Auslander.

In 1971, Noll profited from his Israel connections and published
his *theory of annihilators of linear differential operators*, written in collaboration with H.D. Dombrowski, in the Israel Journal
D'Analyse Mathématique [5.53].

In May and June of 1972, Noll spent six weeks as a visiting professor at the Israel Institute of Technology in Haifa. During this time
he gave a four-week short course on rational mechanics. On May 23,
Noll gave a talk entitled "A Mathematical Theory of Physical Systems with Memory", at the Colloquium in Mathematical Sciences of
the Weizmann Institute of Science in Rehovot. Before the lecture,
Professor Pekeris introduced Noll to the audience as the best American applied mathematician. These flattering words had their price:
Walter Noll had to repeat the colloquium talk at the the Department
of Mathematical Sciences of the Tel-Aviv University in Ramat-Aviv
[5.54]. He wanted to get the Israel honors also as a pure mathematician, so, on June 6, he gave a lecture on the annihilators of linear
differential operators at his host institution in Haifa. He was also invited to give a talk on his theory of inhomogeneities in materially
uniform simple bodies. This took place on June 19, at the Department of Mechanics of Technion and it was warmly appreciated by

the audience. Before his trip to Israel, Noll had finished *a new axiomatization of continuum mechanics*, and he submitted this to the Archive for Rational Mechanics and Analysis. It was such a novel, pioneering, revolutionary work that it was hardly understood to the end even by his teacher C.A. Truesdell, not to mention others.

On March 28–30, 1973, Walter Noll took part in the meeting of the Society for Natural Philosophy at the University of Rochester in New York. He was one of the members of the society which proposed its topic to be "Mathematical Problems Related to Mixtures of Interacting Species".

In June Noll went to Italy, where he gave two lectures entitled "On the Maxima of Functions on Certain Sets of Measures with Applications to Gibb's Thermodynamics" at the summer session on New Variational Techniques in Mathematical Physics in Bressanone.

In 1973, he formulated and submitted an expository paper on the state of art in his thermomechanics to the Archive for Rational Mechanics and Analysis. Although unsatisfied with its content, Noll had to meet the request of C.A. Truesdell to take a definite position among numerous contemporary contributions of his colleagues from the Society for Natural Philosophy. Obviously, he didn't try to receive his due in this paper, and, *with a king's gesture, he abstained from the priority debates*.

In April 1974, Noll gave a seminar talk at the Carnegie-Mellon University. Its topic was "On the Concept of Symmetry of a Physical System" and it repeated his lecture at the Symposium on Symmetry, Similarity, and Group Theoretic Methods in Mechanics in Calgary, Canada. This lecture is an important contribution of Noll's to the *theory of mathematical conceptualization of physical systems*.

The year 1974 brought Noll a new honor: the Springer-Verlag reprinted sixteen of his papers, selected by C.A. Truesdell, as a volume "The Foundations of Mechanics and Thermodynamics: Selected Papers of W. Noll". It is difficult to judge whether this book caused, simultaneously, an unpleasant event: Noll's joint paper with R. Cain, entitled "Convexity, Mixing, Colors, and Quantum Mechanics", was rejected by American physicists. In 1974, Noll's short paper on the second law of thermodynamics in continuum physics was published in the volume "Modern Developments in Thermodynamics".

In 1975, Noll took a serious interest in the possibilities of differential geometry for the conceptualization of physics. In June he gave a talk about his first ideas in this direction at the Seminar on the Foundations of Abstract Differential Geometry at the Carnegie-Mellon University. In October, as a recognized authority, he took part in the meeting of the Society for Natural Philosophy devoted to fluid mechanics, which was held at the Georgia Institute of Technology in Atlanta. In December, he gave a seminar on the applications of differential geometry in rational mechanics at the Mathematics Department of the State University of New York in Buffalo [5.55].

He could hardly concentrate on a comprehensive paper on this topic because his wife became very ill and needed all his attention. From childhood she had suffered from muscle cramps due to an accident. During that year her suffering had become unbearable. They consulted about 30 American physicians, but none of them were able to help. This hopeless condition of permanent, unendurable suffering and her deep love for Walter and their children forced Helga Noll, who was still a strong personality, to try to end her life. Her first suicide attempt failed because Walter was near her. However, in January 1976, he had to fly to San Antonio, Texas, where the annual meeting of the American Mathematical Society was held, and he was unable to prevent her second suicide attempt, which was successful.

On July 5, Walter Noll gave an invited lecture at the Departments of Mathematics and Physics of the Technical University in Berlin. It was intended for novices who had not heard about his mathematical theory of simple materials. In October, as a recognized authority on the theory of dislocations, he took part in the meeting of the Society for Natural Philosophy at the University of South Carolina. His scientific productivity diminished sharply in this year; his only publication, called "The Representation of Monotonous Processes by Exponentials", was an old work from August 1974.

In 1977, Walter Noll was actually able to return to productive mathematical work. There were at least two reasons for this. He found a good friend and an able, sharp-minded, talented collaborator in Juan Jorge Schäffer, one of his colleagues at the Carnegie-Mellon University. Schäffer was a person of encyclopaedic knowledge and had two European Ph.D. degrees. In June 1976, they finished their first joint paper "Orders, Gauge, and Distance in Faceless Linear Cones; with Examples Relevant to Continuum Mechanics and Relativity" {2.42}. It was published in the Archive for Rational Mechanics and Analysis in 1977. The second reason was that he fell in love with a very attractive and clever woman, Mary Strauss. This deep feeling, the only art that Walter could have, changed his life. He began to follow her to the parties, receptions, cocktails and other social gatherings which he had avoided after Helga's death, and could return step-by-step to the normal life of an American university professor. They had very much in common. Both were university professors and had German blood. However, Resie had grown up in the USA and was much better in English and American literature than Noll. She was completely devoted to these subjects. Unlike Noll she was also a good christian. This love became the motor of his everyday life. He began to get pleasure from his lectures. On March 31, 1977, he gave a lecture at the Department of Mathematics of the University of Nothern Illinois in DeKalb, Illinois. It was devoted to his research on orders and metrics in conic spaces. In August, Noll joined a group of his colleagues from the Department of Mathematics at Carnegie-Mellon University, who flew to Brasil to take part in the International Symposium on Continuum Mechanics and Partial Differential Equations, which was held at the Institute of Mathematics of the Federal University of Rio de Janeiro.

The year 1978 was very peaceful for Noll. Only twice did he allow himself to leave Pittsburgh and Resie. In August, he went to Providence, Rhode Island, for the summer meeting of the American Mathematical Society, and, in November, he gave a colloquium for beginners about the concept of simple material at the Stevens Institute of Technology in Hoboken, New York. In 1978, two important papers by Walter Noll appeared. One of them was devoted to the problems in statics of finite elasticity (see {2.44}), while the other, on order-isomorphisms in affine spaces, was written in collaboration with Schäffer and appeared in a very prestigious Italian journal, Annali di Matematica Pura ed Applicata.

On January 4, 1979, Walter Noll and Mary (Resie) T. Strauss were married in Pittsburgh. From that day *Resie Noll devoted herself to her husband and should be credited indirectly for the pearls of mathematical thought that he created thereafter*.

On March 15, Walter Noll gave a colloquium talk on the concept of simple material and its role in elasticity and fluid mechanics at the Westinghouse R & D Center in Pittsburgh. In April he received a great honor to do one of the principal lectures at the Symposium on Albert Einstein at Carnegie-Mellon University. It was called "Einstein and Relativity". This symposium was part of the celebration of A. Einstein's centennial 1879–1979 throughout American universities. Several other distinguished persons were also invited to give lectures there. The main feature of the symposium was the presence of Professor Banesh Hoffmann from Queens College, who was a research collaborator and a friend of Einstein. Hoffmann was the author of one of the popular biographies of Einstein. His symposium contribution was a lecture entitled "Albert Einstein, as the Scientist and the Man". Hoffmann and other prominent American and European scientists evaluated Walter Noll's lecture and his contribution to axiomatical relativity theory very highly. In October Noll received another honor: Sergio De Benedetti invited him to participate on the panel for the 7th Movie on "The Ascent of Man".

In May 1982, Walter Noll was honored to give two main lectures at the Workshop on Categories and Foundations of Continuum Physics, which was held at the State University of New York in Buffalo. However, he submitted only one of them, "Continuum Mechanics and Geometric Integration Theory", for publication in the workshop's proceedings. The materials of the second lecture on the application of differential geometry in continuum mechanics seemed to be only the first step in this direction and their publication would be premature. In September, Noll came to Cornell University where a meeting of the Society for Natural Philosophy was held. Among all his numerous lectures outside Pittsburgh, he didn't forget to give lectures on his achievements at Carnegie-Mellon University. On September 28, he did a seminar talk on the relation of mathematics and physics at the Mellon College of Science. He described in it the role of mathematics as a conceptual tool of physics, using examples from continuum mechanics, thermodynamics and relativity theory. He described his original method of conceptualization of physical theories and he tried to trace its roots in the science of the late 19th and 20th centuries, particularly in the works of G. Peano, R. Dedekind, D. Hilbert, W. Gibbs, and N. Bourbaki.

In 1983, Walter Noll was an honored guest at the Workshop on the Laws and Structure of Continuum Thermomechanics, held at the Institute for Mathematics and its Applications of the University of Minnesota, and at the meeting of the Society for Natural Philosophy at Brown University.

In July 1984, he gave a colloquium talk on the topic "Invariance Principles and Symmetry" at Carnegie-Mellon University. In August Walter and Resie Noll both took sabbatical leave from their university duties for two semesters. They went together to Great Britain, where they intended to stay until the end of the year. For a month Walter Noll got a visiting professorship at the University of Oxford, but most of his time was spent in London. Having known of his arrival in Great Britain, his British friends and admirers asked him to give lectures at their universities. On October 19, he gave an important lecture, "Perspectives in Finite Elasticity", at the School of Mathematics of the University of Bath. On October 12, Professor S. Goldman wrote to Noll and invited him to do a seminar at the Department of Mathematics of the University of Glasgow [5.56]. This lecture took place on November 27. The following day, Walter Noll gave a lecture at the Department of Mathematics of the University of Strathclyde in Glasgow. British listeners were not familiar with his rational mechanics and his achievements were bad news for them. However, the subject of his lecture – the invariance and symmetry principles in physics – was familiar to them and they tried to point out some British priorities in this field to Noll, who was dissatisfied with such a reception and left Glasgow shortly thereafter.

Walter Noll was much better known in Italian scientific circles and he was very happy to see Pisa in January 1985, where he was given a visiting professorship at the university. Resie accompanied him there. In February Noll found free time to visit a conference at the Department of Mathematics of the University of Bologna [5.57]. On March 13, he went to Ferrara, where he did a colloquium talk about his third axiomatic of rational continuum mechanics at the Institute of Mathematics at the local university. After the talk there was no discussion and it looked like a simple social visit. On March 28, he gave a seminar talk about finite elasticity for an audience composed of the staff and graduate students of the University and Polytechnic Institute of Turin. One of the practical results of Walter Noll's stay in Italy was a paper on discontinuous displacements in elasticity, which he wrote in collaboration with a very talented Italian mathematician Paolo Podio-Guidugli. At the end of May, the latter presented it at the Meeting on Finite Thermoelasticity in Rome.

Noll became more and more involved in his lectureship at Carnegie-Mellon University. After his return from Italy, he found a letter at his office from Judith N. Sherwood reminding him that his two lectures at the Mellon College of Science Freshman Science Seminar of the Carnegie-Mellon University were scheduled for September 17 and 19. She wrote there: "The seminar tries to give the students some insight into each area, including an indication of research activities at Carnegie-Mellon University and contributions in the field to the world at large. Each department usually provides two speakers. ... The most successful talks last year were those which used slides, movies, or other visual aids, and which recognized that the audience was quite unsophisticated". Walter Noll announced the topic of these lectures as "Relativistic Time and Fast Interstellar Travel", and they included his original, fascinating calculations of imaginary journeys of spaceships, which illustrated the extraordinary properties of relativistic time.

In 1986, Walter Noll was able to add French mathematicians to the circle of his admirers. In June, he arrived at Nancy, where he had a visiting professorship for a month at the Lorraine National Polytechnic Institute. Noll used this time in full for disseminating his mathematical theories. On June 17, he gave his first lecture, where he demonstrated the applications of geometric measure theory in continuum physics, at the University of Nancy. At his host institution, he gave two lectures. One of them was devoted to his theory of finite elasticity, and the second was on relativistic time and fast interstellar travel. On June 20, he was honored to deliver a lecture on finite elasticity at the University of Pierre and Marie Curie in Paris.

The year 1987 was probably the happiest in his life. First of all, he published the first volume of his fundamental mathematical treatise "Finite-Dimensional Spaces: Algebra, Geometry, and Analysis". Secondly, his friends B.D. Coleman, M. Feinberg and J. Serrin arranged a publication of a volume in his honor at the Springer-Verlag. This book, entitled "Analysis and Thermomechanics: A Collection of Papers Dedicated to W. Noll on his 60th Birthday", included reprints of papers in the volumes 86–97 of the Archive for Rational Mechanics and Analysis. C.A. Truesdell wrote a preface to it.

In January, Walter Noll was invited to give a lecture about the practical foundations of mathematics, which he used to describe his method of conceptualization. In the summer of 1987, Walter and Resie Noll flew to Italy. In June, he was a visiting professor at the University of Pisa. This was also a turning point in Walter Noll's life: he decided that Italian mathematicians were more suitable for collaboration than any others. That summer, he gave two lectures in Italy. The first one was at the Mechanics Seminar of the Mathematics Department of the University of Pisa. It was devoted to his latest research work on the geometry of separation and contact of continuous bodies. It was well received by the audience, and he even got an invitation to repeat it at the University of Rome. In November, Noll made a trip to Morgantown in West Virginia to hold a colloquium on finite elasticity at the West Virginia University.

On March 21–23, 1988, the 25th Anniversary Meeting of the Society for Natural Philosophy was held at the Johns Hopkins University. Among other festivities, the famous American mathematician Walter Noll gave a twenty-minute presentation about edge interactions and surface tension. He also chaired one of the sessions of the meeting. In the after-dinner talk for a banquet of the Society for Natural Philosophy, Professor R. Aris said the following flattering words: "This is an unusual society that here celebrates its first quarter century. Neither seeking to be large in numbers nor exclusive in composition it has (in the words of the preamble) 'nourished specific research aimed at the unity of mathematical and physical science'. By its title, it recognizes the primacy of its goal and so promotes the free intercourse of mathematicians, physicists, chemists and engineers. Because it is aware of its roots in the scientific tradition, it is not obnoxious to the winds of contemporary fashion in the sciences and so is able to take advantage of the latest disciplinary advances without being distorted by them: algorithms may well be developed, but theorems are still proved. Though it publishes nothing, it has indeed, by its meetings and, more generally, through its ethos, 'nourished specific research' that has appeared in the papers of its members. I dare say that if the published papers directly related to talks given at our meetings were collected they would form a very distinguished group of writings. I would like to think too that, thanks to the example of some of the senior members with exacting editorial standards, the decencies of the English language would be as well observed as the demands of mathematical rigor". In 1988, an important joint paper by Walter Noll and Professor E.G. Virga, his most important Italian collaborator, on fit regions and functions of bounded variation was published in the Archive for Rational Mechanics and Analysis.

At the end of May 1989, Noll came to Pisa to give a talk "On the 'Hauptproblem' of Finite Elasticity" at the Meeting on Rational Mechanics and Analysis, dedicated to C.A. Truesdell on his 70th birthday. On June 5, 1989, he gave a seminar talk at the Laboratory of Construction Materials of the Institute of Construction Sciences in Pisa. Its topic was "On Edge Interactions". On June 16, Noll came to Pavia, where he delivered a lecture at the Department of Mathematics of the local university. Its topic was "The Geometry of Separation and Contact of Continuous Bodies" [5.58]. In June, he gave two seminars – on the theory of surface interactions and on geometry of contact and separation of continuous bodies – at the Department of Mathematics of the University of Pisa. At the end of July, Noll and Virga finished an important paper on edge interactions and surface tension, which appeared in the Archive for Rational Mechanics and Analysis in the next year. They dedicated it to B.D. Coleman on the occasion of his 60th birthday.

In November, Noll decided to visit a international meeting of mathematicians in Tbilisi, USSR. He flew from Pittsburgh to the Russian capital Moscow and then to Tbilisi. The meeting was devoted to the 100th birthday of Andria Razmadze, a Georgian mathematician and one of the founders of the University of Tbilisi. Noll brought with him an expository lecture on stability in elasticity. As he remembered, "there was almost no reaction [to it] from the audience" [5.59]. The organizers of the meeting did their best to compensate bad logistics of the conference with traditional Georgian hospitality [5.60]. Walter Noll was very well treated in Tbilisi. A local university chief even gave him two young Georgian mathematicians to accompany him to Moscow to carry his luggage. Noll's visit to Moscow was a social one. He made a sightseeing tour of the city and was invited for dinner with a Russian family.

The year 1991 forms the boundary in the history of Noll's life which cannot be overcome now. It allows us to finish the narrative about it and to move on to further analyse his educational and mathematical works.

6 Mathematics Educator, Researcher and Professor, Walter Noll

Walter Noll considered mathematics to be a special, normed, scientific language. It was to be used instead of ordinary language, when ideas had to be formulated in a precise form. Mathematics and everyday language were very much like two foreign languages, which could for the greater part be translated into each other. Mathematics possessed its own vocabulary and grammar. Very often, one and the same word could have two different meanings in mathematics and ordinary language [6.1]. The term "applied mathematics" was to be understood as the translation of the natural and social sciences into the language of mathematics through a methodical conceptualization. For professional mathematicians, the term "understanding" of a scientific problem was a synonym of its conceptualization with proper mathematical tools. The complete set of such instruments was called a "mathematical frame". Such "mathematical frames" were unique up to their terminology and notation [6.2]. The *process of conceptualization* had to include the following essential steps: (i) proposal of a mathematical frame for a theory, (ii) a conceptualization of the single concepts of the theory, (iii) proposal of a mathematical theory, reproducing the logical structure of the theory, (iv) if such a mathematical theory couldn't be found the mathematical frame should be changed and the procedure repeated [6.3]. It is clear that the process of conceptualization cannot be done by computer and demands, in general cases, more than simply a professional qualification in mathematics. In the history of mathematics of our century, we know only few mathematicians who showed their talent for conceptualization: Georg Hamel, Hans Hermes and Walter Noll belong to them.

According to Walter Noll, one can learn mathematics in two ways: by self-study and by learning from professional mathematicians. In self-study, the main part involves reading mathematical literature. One page of a mathematical book can take much time to read with paper and pencil, and nobody should be reproached for the low speed of reading. However, some parts of mathematics can be learned only from real mathematicians. One should observe them at work and try to understand their actions. The study of mathematics was for Noll similar to the process of how a child learns to speak [6.4]. The central place in the process of learning mathematics belonged to the *concept of proof*. Its role was essential, since it allowed one "to remove any, even unreasonable, doubt". A definition of proof was to be set as an agreement among mathematicians [6.5]. Walter Noll assumed that the whole building of mathematics had only several "key building blocks". The rest followed with their help under the application of the logic of reasoning. In particular, he pointed out three *fundamental mathematical concepts*: relation, group and symmetry. The latter two of these were, at least historically, connected with each other. Noll remarked: "The idea of a 'group' came about as a means of discovering and describing symmetry" [6.6].

He differentiated among the following *three basic types of professional occupation in mathematics*, which could be found in different combinations in every mathematician: *teacher, researcher and professor of mathematics*.

The *main task of a teacher* he considered as conveying a fixed body of knowledge and as focusing students, which should be done in an optimal way. However, the teacher wasn't free in his work, which was based on a textbook and limited by a syllabus. The teacher could be attached to a poor textbook and attracted by the number of "educational gimmicks" in it. He had to pay a great deal of attention to the preparation of homework and pursuit of tests, so as to secure an efficient feedback with his students. The teacher was concerned with his popularity among the staff in his institution and among his students. It very often lead to a lowering of teaching standards, sacrificed to achieve this popularity [6.7]. As a result of his teaching career, Noll was able to formulate his views on the tasks of a teacher and a student in the process of instruction. The teacher was obliged to introduce his students to contemporary abstract mathematics, paying special attention to its most important basic concepts. Noll stressed that a teacher should put emphasis on "mathematics as a conceptual tool rather than merely as a means to express 'quantitative' relations". This meant a transition from elementary "formula- and cookbook-mathematics to real mathematics". The teacher had to give preference in tuition neither to stating nor to solving problems. The contents of tests and exams had to include such problems, which demanded not only "to recite a definition or a theorem from memory". In this sense, they were to be an "open book" for students [6.8]. The *main task of a student* was not simply to memorize and repeat material from the teacher's lectures or from the recommended textbooks of mathematics. He had to learn to think mathematically from his instructor in class. If the student began to think mathematically, the information would stick in his memory by itself. There could be no authority for a mathematical student than his own belief that a certain fact was true. Any material brought to his attention in the process of study was to be suspected and not to be recognized until every doubt was removed. As an instructor of mathematics, Noll rejected the frequently practised system of group study of mathematics and tried to develop competition among his students. However, he didn't show any personal preferences among them in class. Noll attributed the main role in university mathematical education to self-study among students at home or in a mathematical library. This kind of work had to be done with paper and pencil, returning again and again to the passages, which weren't understood at once. The student was to "come up with examples and illustrations for the definitions, propositions, and theorems he read about" [6.9].

There was a considerable difference between the teaching courses, where Walter Noll contributed his own ideas in a significant way, and those which he taught in a more routine manner with the use of textbooks. The latter included such topics as "Statics", "Dynamics", "Strength of Materials", "Higher Mathematics for Science and Engineering Students", "Introduction to Modern Mathematics", "Advanced Calculus", "Linear Algebra", "Computational Linear Algebra", "Algebraic Structures", "Groups, Rings, and Modules", "Topology", "Graph Theory", "Intuitive Geometry" [6.10]. In

Noll's original courses (see Table 6.1), those mathematical students not wanting to learn his original mathematical vocabulary and notation found it difficult to learn anything from him. In class he either dismissed their questions and answers as "meaningless garbage" or explained to them that "real mathematicians would never say that". Answering "wrong questions" or commenting on "improper answers", he used to stress: "If you said that in the test, then you weren't listening in class". He refused to discuss or even to receive ideas which differed from his own. If someone gave an answer to his question that was slightly wrong, even in its wording, Noll reacted very sharply and made him look a fool in front of the whole class. At the same time, when a student answered correctly, Noll said simply "Exactly" and went on with the lecture, avoiding any explanations. He refused to loose time in lectures explaining mistakes. Noll's answer to a correctly posed question in class was usually only a definition or an explanation which might bring the student to the right answer and stimulated him to find it by himself. If the answer of a student wasn't entirely correct, Noll often remarked: "It's clear, O.K., but clumsy". However, he avoided explanations of how to improve the answer. In his original courses, Noll certainly didn't try to follow official course descriptions, and he always openly criticized existing defects and mistakes in university textbooks, listed in such descriptions. In order to compensate the shortcomings of a textbook, Noll's original courses were accompanied by his hand-written notes on theoretical topics. However, he didn't follow the order of materials there, forcing the students to look through them many times in search for the right place. Usually, the notes also didn't contain any examples.

The *main task of the researcher* was to find new results, to discover and to create new knowledge, and his professional career was governed by the "publish-or-perish" rule. The researcher was very sensitive to the issue of priority, and it forced him sometimes to publish premature results. Many researchers had a very narrow field of work, often an obscure and insignificant one, familiar to a small number of mathematicians. They often produced such new results, that could only be understood by themselves, but were easily accepted for publication [6.12].

Table 6.1. Walter Noll's original courses taught at the Carnegie-Mellon University 1956–1993 [6.11]

	Title	Frequency
1.	Mathematical Studies	~ 20
2.	Tensor Analyses (Multidimensional Analysis, Finite-Dimensional Spaces)	> 10
3.	Introduction to Continuum Mechanics (Continuum Mechanics)	~ 10
4.	Differential Geometry	~ 10
5.	Relativity (Special Relativity)	~ 6
6.	Theoretical Mechanics	$1–2$
7.	Finite Elasticity	$1–2$
8.	Measure Theory	$1–2$
9.	Mathematical Logic	$1–2$

Walter Noll has never been a researcher in this sense. However, it is worth presenting his contribution to research mathematics. In 1978, Noll's research work was "concerned with axiomatic foundations of physical theories, in particular continuum mechanics and thermodynamics, with theories of viscoelastic and plastic flow". The spectrum of his research interests also included "differential geometry, especially conceptual clarification of such concepts as affine connections, holonomy groups, and G-structures, applications to general relativity, theory of linear cones, with applications to continuum mechanics, the theory of relativity and quantum mechanics" [6.13]. In 1981–1982, Noll "nearly completed an investigation of the theory of plastic monotonous and viscometric flows", and he "continued to work on the applications of geometric integration theory to the theory of contact interactions and on his theory of transfer structures in differential geometry" [6.14]. In 1983–1984, Noll's research interests included the mathematical foundations of continuum physics and differential geometry [6.15]. In 1985–1986, his efforts were directed at a "continuation of research into the foundations of continuum physics", at the "development of a precise mathematical theory of contact and separation of continuous bodies (with applications to the theory of finite elasticity)", and at the "use of geometric measure theory to obtain a better mathematical structure for describing continuous media" [6.16]. In 1986–1987, Noll obtained a brilliant research result. He wrote about it: "Finally, after a long search, I came up with a good class of 'nice' sets that can serve as regions occupied by continuous bodies" [6.17]. In 1987–1988, he was doing research with a very talented Italian mathematician, E.G. Virga, "on a scheme that would make it possible to include 'edge interactions' in the framework of the general theory of interactions in continuous bodies". Noll was also studying "certain results in differential topology", which could be useful there. Another task of his research was to improve his theory of separation and contact of continuous bodies, which hadn't yet been completed [6.18]. In 1988–1989, he shifted his research interests to the problems of stability in elasticity and the role of quasi-convexity in the problems of stability [6.19]. Over the next two semesters at the Carnegie-Mellon University, Noll continued his research in elastic stability and resumed his studies of geometry of contact and separation of continuous bodies [6.20]. In about 1989, he formulated his research plans for the rest of his professorship at the Carnegie-Mellon University (see Table 6.2). A general evaluation of Walter Noll's research activities can be found in the document "The Mathematical Studies Program: Remarks by Walter Noll". It consisted, as he wrote, mainly in the "conceptual analysis of continuum mechanics and thermodynamics and of the theory of relativity, using the language of modern mathematics".

The third type of professional activities in mathematics was for Noll to be a *professor of mathematics*, a person "professing his subject: mathematics". He included himself among this type of professional. A professor had to think, to discuss and to lecture his subject, which he "lived". He never followed a textbook or a syllabus. "He recreated the subject in his mind each time he lectured on it", and thus he could change the art and order of the course materials, to include or exclude them. The *main teaching task of a professor* was "to find a new approach to and better insight into the subject of his course". A professor didn't pay much attention to homework, tests or grades, considering them to be a low-level job for his teaching assistants. Noll saw the role of a professor

Table 6.2. The research plans of Walter Noll after 1989 [6.21]

Title	Comments
1. The Geometry of Separation, Contact and Transition in Continuous Bodies	These processes have never before been described in precise mathematical terms; such a description may lead to a better understanding of the physics of cracking, glueing, cavitation, phase transition, and similar phenomena.
2. Contact Interactions	Concept of an abstract contact interaction can be applied to both mechanics (forces and torques) and thermodynamics (heat transfer); many questions in the mathematical theory of contact interactions still remain open.
3. Elasticity	In finite elasticity there remain important conceptual questions concerning the physical interpretation of the mathematical results and the basic constitutive functions.
4. Plastic, Monotonous and Viscometric Flows	Application of the new concept of simple material to such flows.
5. Time Event-Worlds	The mathematical concept of a "time event-world" can provide an infra-structure not only for classical and relativistic time-concepts, but perhaps also for at yet undeveloped time-concepts appropriate to relativistic quantum worlds; the proposed research may help make such radical new time-concepts possible.

in mathematics as understanding, gaining insight into, judging the significance of and organizing old knowledge.

The teacher of mathematics depended on the work of the professor in order to keep up with the modern context in the field. The researcher was also a minion of his, since it was the professor who examined research works critically, sorted them and fitted them into a coherent framework. For the professor, there was no priority problem: "before going into print he let his ideas ripen". The *main task of a professor* in research was to fit into a more general context (which was first to be found), to unify and to simplify the "undigested and ill-understood new results". A professor was a mediator between the researcher and the teacher. He "got no recognition for writing elementary textbooks" and he was only pleased to know about positive student evaluations of his teaching [6.22].

Noll's academic career wasn't decorated enough with honors and rewards. He wrote about it: "In the Faculty Handbook of this University (Carnegie-Mellon), under 'Criteria for Faculty Appointments', I can find almost nothing that relates to professing a subject in the sense described above. Such professing of a subject rarely gets much recognition. Most of the rewards in academia go to those who excel in research or in teaching" [6.23]. Noll stressed that "he had spent most of his professional life neither with 'teaching' nor with 'research'" [6.24]. The spring semester of 1993 was the last in Noll's teaching career at the Carnegie-Mellon University. One of the main reasons for this was the lack of popularity for Noll's mathematical courses among students. He wrote about this with bitterness: "Most students do not know the difference between a teacher and a professor. They expect to be treated the same way in college as they were treated in high school. They do not know that, in college, they should be their own teachers" [6.25].

It is worth presenting the perception of Noll's instruction of mathematics by his students which can be found in the available faculty-course evaluations of his lectures at the Carnegie-Mellon University. A minority of students appreciated Noll's statement that the professor should never follow a textbook or a syllabus and understood this as "motivation to seek out readings in the library". The overhelming majority of them reacted negatively to the lack of textbooks, especially in the courses of linear algebra (this was before 1987, when Walter Noll's treatise on finite-dimensional spaces {1.5} was published). Without a textbook, these students described Walter Noll's courses as lacking reference points, any class organization and a definite structure. The hand-written notes to his courses were evaluated by them as very sketchy, too abstract, lacking examples, solutions and proof of theorems, having no connection with tests, overly wordy and dependent on a special symbology, invented by Noll himself. Noll's attempt to recommend a textbook of linear algebra in the fall semester of 1985 didn't change the situation, since he didn't follow it. Students wrote in the faculty course evaluations: "The book is no help (whatsoever). We have nothing to fall back on if we don't understand the notes". The students found it confusing when Noll said in class, on the one hand, that they should read the textbook, but, on the other hand, that it was bad and should be neglected. The minority of students who already had some experience of Noll's teaching, found it alright. One of them remarked: "Professor Noll tells us what parts of the book pertain to our course and gives out handouts for other materials". The American students were very formal in their approach to the evaluation of Noll's lectures. When he moved away from the officially existing description of the linear algebra course and emphasized some basic theory and proof instead, many students got the feeling that they weren't getting what they should. Some of them abandoned the course and tried to learn linear algebra on their own. They wrote in particular: "The material he (Walter Noll) covered was barely 30% of that listed in the catalog. He expects us to know things in tests we have not gone over at all, and even when they are in the book, he expects us to do them *his way*". One of Noll's successes was his participation in a special honors program, "Mathematical Studies", devoted to the exposition of basic mathematics. These lectures were evaluated by students as excellent and very well organized.

Noll's statements that the professor was free (i) to recreate "the subject in his mind each time he lectured on it", (ii) to change the art and order of the course material, (iii) to exclude or include it at any time, and their application in the teaching practice were criticized by his students. His most severe critics were again his students at the linear algebra classes. They accused him of being too theoretical, paying too much attention in theory to trivial details of form and notation. Some of them found his courses to have no order and to be lacking in landmarks. The size of course remained unknown to them. The American students were especially worried that the content of Noll's lectures hadn't always corresponded to the course description. He avoided numerous topics which, according to the description, should have been taught. Students couldn't understand how the course structure could be flexible, lacking a fixed direction. As a strong point of Noll's lectures, they pointed out his emphasis on the "concepts and ideas behind mathematics – not just recipes to memorize". In the fall semester of 1985, Noll attempted to explain to his students that the linear algebra course was nothing more than a conglomerate of parts of the other mathematical disciplines. This led them to confusion. They understood it wrongly as an attempt to defend his "faulty lectures", which had no structure, no plan and no direction. They went as far as to propose incorporating the linear algebra course into other mathematical courses and thus eliminating it completely. The linear algebra students wrote that Noll's course was so original that it couldn't serve any purpose in other classes at the Carnegie-Mellon University or in real life. Noll tried to teach them serious mathematics, which was for him a synonym of theoretical mathematics, but it took a long time for his students to understand it.

Noll wrote that his professor hadn't paid much attention to home-work, tests and grades. As it appears from faculty-course evalua-tions, Noll has never been explicit on grading of class performance, homework or tests. He avoided a fixed grading scheme, so students could hardly ascertain how they were doing and were forced to work the whole time (what Noll actually wanted to achieve). Students called it the "I-have-no-grading-scheme" policy, which they con-sidered unfair. Noll wanted to estimate the performance of students on an individual basis. However, it must have been very difficult at times due to their large number (up to 145 students in class). In or-der to overcome this difficulty, he seemed to change the grade level so as to allow the majority of students to pass. He gave homework of different levels of difficulty. Many of the problems were evalu-ated by students to be either puzzling or too simple and insignificant in comparison with those in tests. Noll posed no midway problems. He used to invent homework problems "off the top of his head" in class. Most of the homework were proofs, generally avoiding ex-amples and applications of the class materials. The solutions were evaluated by Noll as "positive" or "negative". The student himself was left to think about the correct solution. Noll avoided giving any credit for the homework in class. He announced tests one week in advance, and those students who didn't learn the material by self-study and weren't attentive in class, were in great difficulties. They didn't know what to learn or to revise for the tests.

During his teaching career at the Department of Mathematics at Carnegie-Mellon University, Noll advised four *graduate students*. The first of them was Michael Summers, who wrote his diploma under Noll's scientific advice between 1978 and 1980. He graduated from the Mellon College of Science in June 1980. In the fall semester of 1981, Noll had another graduate student, Edward C. Pervin, who left him soon after changing department. In the fall semester of 1985, Sandy Samelson obtained the grade of master of applied mathematics, under the guidance of Noll. In the spring semester of 1987, Noll began to advise a talented graduate student, Vincent J. Matsko. Matsko was born in Sharon, Pennsylvania, on August 10, 1964. He studied mathematics at the Carnegie-Mellon University, and, in 1986, obtained a bachelor's degree there. Under Noll's scientific guidance, Matsko received a master's degree in applied mathematics there in 1989. Matsko's major mathematical interests were in the field of modern geometry. In 1993, he also obtained a doctorate of arts under Noll's guidance. The content of his dissertation was based on Noll's course of relativity theory.

The first of Noll's *doctorate students* was Wayne Van Buren who defended a Ph.D. thesis at the Carnegie-Mellon University in 1968. Its was called "On the Existence and Uniqueness of Solutions to Boundary Value Problems in Finite Elasticity". Van Buren's scientific career was connected further with the Department of Materials and Systems R & D of the Westinghouse Research Laboratories in Pittsburgh. Van Buren's dissertation and two known Westinghouse research reports, written in 1967 and 1968, were devoted to the theory of finite deformations in elasticity. He managed to clarify and generalize certain results of F. Stoppelli on the local existence and uniqueness of solutions of finite and infinitesimal traction boundary value problems, including place and mixed problems [6.26].

In April 1978, another of Walter Noll's pupils obtained a doctorate of philosophy at the Carnegie-Mellon University. His name was Ray Edison Artz, Jr. The dissertation was entitled "An Axiomatic Foundation for Finite-Dimensional Quantum Theory". In its abstract, Artz wrote: "A *sliced cone* is a convex linear cone equipped with a prescribed strictly positive linear functional. In quantum theory (and in more general theories of measurement), members of the cone and the prescribed functional can be interpreted, respectively, as beams of particles and as a device which counts the numbers of particles per unit time carried by those beams. In 1969, Mielnik (Commun. Math. Phys. 15, 1–46) pointed out that complex Hilbert spaces play their role in quantum mechanics only through the intrinsic geometry of their sliced cones of trace-class positive Hermitian operators (equipped with the trace functionals). This suggested that a physically direct quantum theory could be developed within the theory of sliced cones, but there remained the problem of finding an *intrinsic* characterization of those sliced cones which are applicable (that is, which are isomorphic to sliced cones of trace-class positive Hermitian operators on complex Hilbert spaces). The present ... (dissertation) solves this problem in the case where the cones are finite dimensional. In doing so, it provides considerable insight into why complex Hilbert spaces (rather than real or quaternionic Hilbert spaces) are relevant to quantum mechanics".

In 1988, a Chinese mathematician, Sea-Mean Chiou, asked Walter Noll to be the advisor of his Ph.D. dissertation. Noll agreed, probably influenced by the recent murder of a Chinese colleague in Taiwan, executed on charges based on the spying activities of other Chinese students on campus at the Carnegie-Mellon University in Pittsburgh [6.27]. Chiou was born in the city of Taipei, Taiwan R.O.C., on January 2, 1956. He obtained a bachelor's degree at the Tung-Hai University in 1978 and a master's degree in mathematics at the Lamar University, Texas, in 1985. In 1992, Chiou defended a Ph.D. dissertation on differential geometry at the Carnegie-Mellon University. Its content was based on the lectures of Noll on differential geometry.

Noll assisted in five other doctoral dissertations.

In October 1963, he advised on Lincoln E. Bragg's thesis "On Relativistic Worldlines and Motions, and on Non-Sentient Response". This included some new results on pseudo-euclidean space and space-time, topology and worldlines, signal emission and reception times, smoothness of worldlines and signal time functions, a signal preserving map, the class of signal preserving maps, apparently identical motions, and the non-sentient response [6.28].

In 1970, Noll served as advisor on a thesis presented by Michel Lucius to the Faculty of Science of the University of Nancy, France, to obtain the degree of doctor of mathematical sciences. Its subject was "Etude d'une Équation Différentielle Fonctionnelle Non Linéaire Régissant les Mouvements Instationnaires de Fluides Simples". Lucius gave a high evaluation of Noll's role as advisor on his thesis. He wrote: "The present work was realized under the guidance of Prof. Walter Noll, professor of mathematics at the Carnegie-Mellon University, Pittsburgh (USA). We wish to express here our deepest gratitude to him for the innumerable pieces of advice which he gave us so generously". This dissertation was devoted to the principle of fading memory, including a new method of solving of the functional-differential equation for simple fluids and an original proof of the existence, uniqueness and asymptotic behavior of the solution and a determination of relaxation functions of simple fluids.

In 1972, Rolf Zuber received a doctorate of natural sciences at the Swiss Technical University in Zurich, Switzerland. The subject of his thesis was "Einfache Körper mit konstanter Inhomogenität". Noll's role in it received a warm evaluation from Zuber, who wrote: "The present work came into being during the period of my assistantship at the Carnegie-Mellon University in Pittsburgh and at the Swiss Technical University in Zurich. . . . I received stimulus for my work from Prof. W. Noll in Pittsburgh. I thank him cordially for this and also for many fruitful discussions". This thesis included some interesting results on material bodies, simple bodies, material uniform bodies and symmetry, homogeneous bodies and inhomogeneity, most general constants of inhomogeneity, special and most general reference function with constant plane inhomogeneity, controllable configurations and deformations.

In 1978, Noll served as a referee for a Ph.D. thesis in mathematical physics, submitted by Paul A. Carlson to the Department of Mathematics at the University of Alberta in Edmonton. Its subject was "Relativistic Uniform Continua".

During his stay in France in the summer of 1986, Noll participated in the examination committee for the grade of "Doctorat d'Etat" at the National Polytechnic Institute of Lorraine. The degree was conferred upon Didier Bernardin.

Noll took an active part in the administration of the Carnegie-Mellon University. In the fall semester of 1980, he was elected to the Committee on the Undergraduate Honors Curriculum of the Department of Mathematics of the Carnegie-Mellon University. In the fall semester of 1981, he was appointed to be chairman of the Mellon College of Science (MCS) Ad Hoc Committee on Tenured Promotions and Reappointments. He occupied this position until the fall semester of 1985. In the fall semester of 1987, Noll became a member of the Faculty Senate and resumed his participation – as an ordinary member – at the MCS Ad Hoc Committee on Tenured Promotions and Reappointments. In the fall semester of 1988, having kept only his position at the Faculty Senate, Noll became engaged in the activities of the Educational Affairs Council and at the UEC Task Force on Faculty Course Evaluations. He and his family took an active part in the experimental program of the Student-Alumni Association, which sponsored home gatherings of staff and students in Pittsburgh. On several occasions, they invited to their house four students from various departments of the Carnegie-Mellon University and also three couples: one alumnus couple and two faculty couples. The purpose of these Sunday gatherings at Noll's was "a good conversation and an exchange of ideas among people with different interests, backgrounds and ages" [6.29].

From 1962, Noll was involved in editorial work as a referee of the following international and American scientific journals: A.I.Ch.E. Journal, Biorheology, Foundations of Physics, Indiana University Mathematics Journal, International Journal of Engineering Science, International Journal of Solids and Structures, Journal of Applied Mechanics of ASME, Journal of the Franklin Institute, Journal of Mathematical Physics, Journal of Nonlinear Analysis (Theory, Methods, and Applications, JNA-TMA), Journal of the Society for Industrial and Applied Mathematics, Journal of the Society of Rheology, Mechanics (Research Communications), Philosophy of Science (Official Journal of the Philosophy of Science Association),

The Physical Review, The Physics of Fluids, Quarterly of Applied Mathematics, Rheologica Acta (An International Journal of Rheology) and many others. Between 1959 and 1976, Noll served as a member of the Editorial Board of the Archive for Rational Mechanics and Analysis (Volumes 3–63), published by the Springer-Verlag [6.30]. From October 1971 to March 1973, he was a co-editor of the series "Springer Tracts in Natural Philosophy" [6.31]. As a referee of professional journals, Noll often rejected offers to judge papers which didn't correspond to his mathematical qualifications or which dealt with the old system of continuum mechanics [6.32]. At the same time, he used to give negative references for those articles which didn't have a clear physical application [6.33].

Unlike many of his ambitious and celebrated colleagues, Walter Noll has become a member of only two professional associations: the American Mathematical Society and the Society for Natural Philosophy [6.34].

Fig. 1. Walter Noll, Licencié ès Sciences of the University
of Paris, in 1950

Fig. 2. Walter Noll, Scientific Assistant at TU Berlin, in 1952

Fig. 3. Photo of the participants of the First Meeting of the Society for Natural Philosophy, March 25–26, 1963, at the Johns Hopkins University (from left to right): first row – L.M. Milne-Thompson, C.A. Truesdell, B.D. Coleman, J.B. Serrin, J.L. Ericksen, H. Markovitz, E. Sternberg, W. Noll, H. Grad, R.A. Toupin, R.S. Rivlin; second row – S. Corrsin, G.S. Benton, O.W. Dillon, M.F. Beatty, W.H. Pell, A. Ghaffari, A.J.A. Morgan, E.H. Lee, C.C. Hsiao, V.K. Stokes, D.C. Leigh, J.L. Finck; third row – E. Saibel, R.E Green, J.H. Suckling, C.J. Jeffus, H. Zorski, J.H. Weiner, R.C. Wanta, I. Herrera, M.E. Gurtin, W.T. Sanders, M.E. Rubenstein; fourth row – F. Grün, A.B. Metzner, E. Ikenberry, W.R. Schowalter, W.N. Sharpe, W.J. Gillich, P.J. Blatz, G. Lianis, Z. Karni, G.L. Hand, W. Jaunzemis, S.C. Cowin; fifth row – V.J. Mizel, P.M. Naghdi, E.A. Fox, H.F. Tiersten, W.D. Collins, L.E. Payne, J.H. Bramble, P. Mann-Nachbar, W.R. Foster, J.P. Archie, W.S. Ament, J.A. Simmons; sixth row – J.L. White, A.C. Eringen, Y.T. Huang, A.W. Saenz, B. Bernstein, R.B. Pond, G.L. Filbey, F.J. Lockett, S.C. Hunter, R.N. Thurston, M.V. Morkovin, R.J. Eichelberger; seventh row – O.M. Phillips, J.L. Lumley, P.D. Kelly, A. Phillips, J.L. Thompson, A.C. Pipkin, J.A. Lewis, J. Sperrazza, Mrs. C. Truesdell, R. Finn

Fig. 4. Bernard D. Coleman, Noll's major collaborator
in 1958–66

Fig. 5. Heinz Dieter Dombrowski, Noll's major collaborator in 1967–71

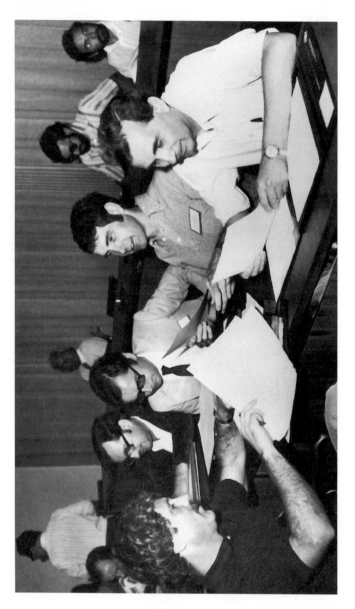

Fig. 6. At the International Symposium on Continuum Mechanics and Partial Differential Equations, August 1–5, 1977, in Rio de Janeiro (from left to right): M.E. Gurtin, B.D. Coleman, P. Podio, W.O. Williams, W. Noll

Fig. 7. Epifanio G. Virga, Noll's major collaborator since 1985

Fig. 8. Photo of the participants of the 25th Anniversary Meeting of the Society for Natural Philosophy, March 21–23, 1988, at the Johns Hopkins University (from left to right): first row – M.E. Gurtin, R.G. Muncaster, M.F. Beatty, Jr., W. Noll, R.M. Bowen, B.D. Coleman, J. Serrin, C.-C. Wang, S. Spector; second row – M. Feinberg, R. Fosdick, R. Batra, A.W. Marris, V.J. Mizel, C.A. Truesdell, W.O. Williams, S. Passman, G.S. Benton, W.N. Sharpe, Jr.; third row – E.G. Virga, R. Aris, P. Holmes, P. Podio-Guidugli, J.F. Bell, G. Capriz, L.C. Tartar, W. Ziemer, C. Amick, J.W. Nunziato, C. Dafermos; fourth row – S. Antman, M. Ferrari, K.B. Knami, E. MacMillan, G. Del Piero, F. Davi, J.W. Walter, Jr., L. Cesari, L.C. Martins, D. Fisher, S. Messaoudi, M. Slemrod; fifth row – E.K. Walsh, I. Stakgold, P.C. Fife, M.E. Mear, M. Epstein, R. Segev, P. Ponte-Castañeda, X. Liu, Y. Zhang, D.S. Bridge, M. Tarabek, K. Wang, M. Massoudi; sixth row – A. Nachman, R.C. Rogers, J.H. Maddocks, R. James, G.F. Abatt, D.E. Carlson, G. MacSithigh, M.J. Scheidler, R.S. Marlow, J.F. Pierce, T.J. Burns, H.C. Simpson; seventh row – S.N. Ganeriwala, G. Batra, J. Simo, L. Davison, J. Greenberg, P.H. Rabinowitz, K.R. Rajagopal, T.W. Wright, A.E. Tzavaras, J.D. Humphrey, C.-S. Man; eighth row – Y.-C. Chen, H.S. Sellers, S. Adeleke, H. Hattori

Fig. 9. Walter Noll with his wife, Resie, in 1989

7   Annihilators of Linear Differential Operators

W alter Noll published two works related to the theory of anni- hilators of linear differential operators. The first publication, entitled "Quasi-Invertibility in a Staircase Diagram", was finished in January 1969. It contained a theorem, "which was needed in an in- vestigation of annihilators of differential operators, but could have other applications" [7.1]. The second paper was devoted to a gen- eral theory of annihilators of linear differential operators. Noll origi- nally planned to publish it in the Proceedings of the American Math- ematical Society, but then, for unknown reasons, submitted it at the end of July 1969 to the Journal D'Analyse Mathématique, issued in Jerusalem, Israel [7.2]. The theory of annihilators did not belong to Noll alone. It was created by him in collaboration with a young talented German mathematician Heinz Dieter Dombrowski. More- over, the idea of the work originated from the habilitation script of Dombrowski's, "Eine Charakterisierung der Differential Opera- toren", published in 1966 [7.3].

H.D. Dombrowski was born on November 3, 1936, in Kallenau, Germany. He received a professional education as a mathematician and defended a Ph.D. thesis on the topic "Fastautomorphe Funktio- nen zweiten Grades" in 1962 at the famous University of Göttingen [7.4]. Between 1963 and 1966, Dombrowski worked intensively on the mathematical foundations of physics. In collaboration with Klaus Horneffer, he published a paper in 1964 on the axiomatiza- tion of the concept of a physical system and another on differential geometry, related to the Galilean Principle of Relativity [7.5]. Dom- browski and Horneffer devoted also a paper to a generalization of the Galilean Principle of Relativity through the introduction of an arbitrary coordinate system into the space-time continuum [7.6].

Noll met Dombrowski at the Workshop on the Foundations of Physics held in Oberwolfach, West Germany, in June–July 1966. Noll liked Dombrowski's report and he got to know Dombrowski personally. After a conversation, Dombrowski considered him as a "very interested and susceptible partner to speak to". Their scientific interests seemed to coincide, especially in axiomatical physics and space-time theories. Dombrowski remembered: "Apparently, he (Walter Noll) liked my way of dealing with fundamental physical problems. It had some similarity with his own approach, although we used completely different ... mathematical methods. But exactly this different initial position seemed to appeal to Noll" [7.7]. Without hesitation, Noll invited Dombrowski to come for a one-year-stay to the Department of Mathematics of the Carnegie-Mellon University. With Noll's assistance, Dombrowski applied and succeeded in getting a Senior Foreign Scientist Fellowship from the National Science Foundation of the USA [7.8]. Dombrowski arrived in Pittsburgh in October 1967. He remembered: "The year from October 1967 to September 1968 in Walter Noll's team at the Carnegie-Mellon University was for me a wonderful, stimulating, and enjoyable time" [7.9].

Noll thought he had found a partner, with whom he could work on the mathematical foundations of physics. However, it turned out very soon that their views of this subject were different, and Noll didn't intend to change his views. Another difficulty was of technical character: their contact time at the Carnegie-Mellon University was only several hours a week; too short for a real collaboration [7.10]. At the same time, Dombrowski could enjoy the hospitality of Noll and his colleagues at the Department of Mathematics in full. He took part in all the parties, cocktails, receptions and dinners which were typical of the social life of the teaching staff at the Carnegie-Mellon University. However, the approaching end of his stay in September 1968 troubled Dombrowski, who had to give an account of its practical results to the National Science Foundation. He decided to produce a joint publication with Noll in pure mathematics, in which they both seemed to have common interests. Dombrowski offered Noll an algebraic topic on annihilators of linear differential operators, which was one of the possible continuations of the ideas of Dombrowski's habilitation work of 1966 [7.11].

Noll's contribution to the theory of annihilators of linear differential operators was highly regarded by Dombrowski, who wrote about it in 1991: "But of the real work that went into the publication, Noll contributed just as much as I did. He familiarized himself amazingly quickly with this algebraic treatment of linear differential operators and finally overtook me as shown by his separate publication on the staircase diagram" [7.12]. After his return to West Germany, Dombrowski continued to work on the joint manuscript, exchanging results with Noll by post until about 1971. In 1969, Noll communicated a paper of Dombrowski's on simultaneous measurements of incompatible observables in non-relativistic quantum physics to the Archive for Rational Mechanics and Analysis. After about 1971, all contacts between the two mathematicians slowly ceased [7.13].

Let us briefly consider the general algebraic theory of annihilators of linear differential operators, built by Noll and Dombrowski [7.14]. Their approach was constructive. Based on some results of Dombrowski and J.M. Boardman's, the authors put the question which predetermined the structure and content of their research: "Under which conditions does a differential operator admit a generating annihilator, and how can it be determined if it exists?" [7.15]. At the first step, Noll and Dombrowski introduced a concept of a differential operator and an operation of product on the set of such operators, in respect to which there existed a zero differential operator. If one took a differential operator from a fixed subset it would be possible to build a set of differential operators from the given operator by multiplying it with some differential operators from another fixed set. Then, the operators of the first set were called "generating" differential operators. For a given differential operator, the annihilator was defined as another differential operator, whose product with the given operator was equal to the zero differential operator. In order to find a generating annihilator, Noll and Dombrowski turned their attention to two of its known characteristics: characteristic polymorphism and curvature. The second characteristic imposed the main limitation on the Noll-Dombrowski's theory: "It applied only when the curvatures k_s (of a differential operator) vanished for large s" [7.16]. The principal idea of how to achieve the purpose of the study was to pass from analysis of differential operators to that of their polymorphisms. Noll and Dombrowski explained this in {2.36}: "Polymorphisms are special homomorphisms, and as such they are more accessible to algebraic treatment than differential operators. The composition of differential operators suggests a certain composition for polymorphisms ... , which can be used to define annihilators of a polymorphism in a way similar as for differential operators. The problem of finding a generating annihilator of a given differential operator has a counterpart in the theory of polymorphisms. The two problems are not only analogous, they are also related via the process of taking characteristic polymorphisms .... The main result (of our theory) is Theorem 1 .... Roughly speaking, it enables one to find a generating annihilator of differential operator f provided one knows a generating annihilator of the characteristic polymorphism of f and provided all curvatres of f vanish" (pp. 209–210). Noll and Dombrowski added to the theoretical part of the paper a comprehensive section devoted to its applications to five "standard examples of differential operators", including the one characteristic for infinitesimal elasticity [7.17].

8 Finite-Dimensional Spaces

In 1987, the first volume of Walter Noll's fundamental treatise on finite-dimensional spaces was published. He tried to present a brief sketch of its genesis in the introduction. Unfortunately, as in several other cases, Noll's historical exposition is defective. It is the primary aim here to reconstruct a more probable history of its creation.

During the spring semester of 1962, Noll spent much of his time, when not teaching, preparing notes on tensor analysis. In the following semester he moved to Baltimore as a visiting professor at the Johns Hopkins University. He remembered: "I never gave any course at all when I was at Johns Hopkins. C.-C. Wang was a graduate student there and he took what is called a 'reading course' at American universities: I gave him my handwritten notes (on tensor analysis) and asked him to read and study them, and then we had one-on-one discussions from time to time" [8.1]. C.-C. Wang arranged his records of talks with Noll and Noll's notes into a manuscript which appeared as a department issue in October 1963 [8.2].

This work {1.1} included six chapters, divided into 22 paragraphs. It began with definitions of the concepts of vector space and its dual. Then, the concepts of inner product space and linear transformation were introduced. In the next chapter the conceptual analysis was expanded to the Cartesian product of spaces. New expositions of Grassman algebra and vector analysis were added to these materials. The last four paragraphs were devoted to some applications of these abstract mathematical conceptual tools in continuum mechanics. Noll defined in {1.1} the velocity and acceleration fields of a material particle in a three-dimensional Euclidean point space ε. Then, a modified definition of a continuous body (in comparison

with Noll's first axiomatization of continuum mechanics presented
in Section 8.1 of the book) was given: "A set B of elements Z is
called a *continuous body* if there exists a family Φ of mappings from
B to a domain in ε with the following properties: (i) Every element k
in Φ is one-to-one; (ii) If k, λ are elements in Φ then the composition
$k \circ \lambda$ is a deformation from $k(B)$ onto $\lambda(B)$; (iii) If k is in Φ and if $\underline{k}$
is a deformation of $k(B)$ then the composition $\underline{k} \circ k$ is in Φ" [8.3]. A
point valued field which mapped a domain in ε into another domain
in ε was called a *deformation* if the mapping was one-to-one and
continuous in both directions (see {1.1}, p. 72). Noll also introduced
the following familiar concepts of kinematics of continuous bodies:
body configuration, steady motion, stream line, velocity potential,
path line, reference configuration, deformation gradient of the motion relative to the reference configuration, right and left stretch tensor of the motion, rotation tensor of the motion, Cauchy-Green tensor, the deformation gradient at a time moment relative to another
time moment, spin tensor, stretching tensor, velocity gradient of the
motion, irrotational motion, and potential flow [8.4].

Noll proceeded to improve his notes on tensor analysis and they
served as a basis for the section on tensor functions included in the
fundamental treatise of C.A. Truesdell and Noll on non-linear field
theories of mechanics, which was published in 1965 [8.5]. At about
that time, Noll's notes in different forms became widely known
and something like a nationally acknowledged standard source of
"mathematical concepts essential to continuum physics" [8.6]. He
included an advanced version of the notes as an appendix to his joint
monograph with B.D. Coleman and H. Markovitz on the viscometric
flows of non-Newtonian fluids {1.3}, which appeared in 1966 [8.7].
This appendix comprised the following concepts: vector and inner
product spaces, linear form, Euclidean (point) space, tensor, tensor
algebra, derivative, gradient of a point, vector function, tensor function, and deformation. In order to increase the practical value of the
monograph, Noll included in the appendix the concept of a coordinate system and described the mostly used Cartesian, cylindrical and
spherical coordinates [8.8]. C.A. Truesdell used a version of Noll's
notes on tensor analysis to produce an appendix with useful concepts
and results of algebra, geometry and calculus, which he included in

his most famous course of rational continuum mechanics, published in 1977 and, in revised form, in 1991 [8.9].

The preliminary version of Noll's first volume of the treatise on finite-dimensional spaces included 138 type-written pages, subdivided into six chapters and 35 paragraphs. They comprised, correspondingly, the following mathematical frames: theories of inner product spaces, tensors, Euclidean point-spaces, differentiation, coordinates, and the structure of linear transformations. From 1973, it took Noll more than ten years of hard work to prepare this preliminary text and to test its quality in teaching practice [8.10]. Between approximately 1983 and 1987, he made the last revision of the text of the first volume. It contained 393 pages, ten chapters, subdivided into 95 paragraphs. The first two chapters were intended for the layman with no professional training in mathematics. Noll included in them the following materials: basic set theory, algebraic structures, linear algebra and functional analysis in one dimension [8.11]. Each chapter of the treatise included a list of problems, which weren't present in the preliminary version, and each paragraph ended with notes explaining the notations and terms used in it. Especially for this treatise, Noll elaborated an original system of notations and a new terminology for the finite-dimensional spaces [8.12]. Less than ten paragraphs of the text of 1987 repeated those of the preliminary version. The order of materials was considerably changed. Three chapters of the preliminary version retained their status, but were notably improved. The main body of Noll's first volume of the treatise included the following mathematical frames: theories of duality, bilinearity, flat spaces, inner-product and Euclidean spaces, coordinate systems, general lineons, elements of topology, differential calculus and spectral theory.

Between 1984 and 1987, Noll wrote several chapters to serve as a basis for the second volume of the treatise. They included materials about volume integrals, multilinearity, differentiable manifolds and manifolds in an Euclidean space [8.13].

9 Foundations of Rational Continuum Mechanics

Before the first axiomatization attempt of continuum mechanics, Walter Noll studied throughly the comprehensive work of G. Hamel, "Die Axiome der Mechanik" (1927), and the guidelines of David Hilbert on the axiomatization of mechanics [9.1].

9.1 First Axiomatic of Rational Continuum Mechanics

The first work of Noll, containing an axiomatic of continuum mechanics, appeared in 1957 and was entitled "The Foundations of Classic Mechanics in the Light of Recent Advances in Continuum Mechanics" [9.2]. Its purposes were: (i) to outline an axiomatic scheme for continuum mechanics; (ii) to conceptualize continuum mechanics on "the same level of rigor and clarity as was then customary in pure mathematics" [9.3]. This axiomatic had two defects: (i) it wasn't universal and was, perhaps, too special: "it did not cover concentrated forces, contact couples and body couples, sliding, impact, rupture, and other discontinuities, singularities, and degeneracies"; (ii) it didn't include the thermodynamical processes [9.4]. Walter Noll divided his axioms in three groups:

Axioms of Continuum Body. The first group was to make clear the concept of body in continuum mechanics. The mathematical framework for a theory of bodies was that of differentiable manifolds and of the measure theory. Noll defined a continuum body as a triple $\langle B, \Phi, m \rangle$, where B was an arbitrary set, Φ was a set of mappings from B into E, a three-dimensional Euclidean point-space, m was a function, defined on the subsets of B, into the set of real numbers. B was called a set of particles of a continuum body. Φ was called a set of configurations of the body in space. m was called the mass distribution inside the body. The following axioms were postulated in this group:

N1X: Every mapping $\phi \in \Phi$ is one-to-one [9.5].

N2X: For each $\phi \in \Phi$, the image $B' = \phi(B)$ is a region in the space E, a region being defined as a compact set with piecewise smooth boundaries [9.6].

N3X: If $\phi \in \Phi$ and $\psi \in \Phi$, then the mapping $\chi = \psi\phi^{-1}$ of $\phi(B)$ onto $\psi(B)$ can be extended to a smooth homeomorphism of E onto itself.

N4X: If χ is a smooth homeomorphism of E onto itself and if $\phi \in \Phi$, then also $\chi\phi \in \Phi$.

N5X: m is a non-negative measure, defined for all Borel subsets G of B [9.7].

N6X: For each $\phi \in \Phi$, the measure $\mu_\phi = m\phi^{-1}$ induced by m on the region $B' = \phi(B)$ in space is absolutely continuous relative to the Lebésque measure in B. Hence it has a density ρ_ϕ so that

$$m(G) = \int_{\phi(G)} \rho_\phi(x)\mathrm{d}V \ .$$

N7X: For each $\phi \in \Phi$ the density ρ_ϕ is positive and bounded [9.8].

Axioms of Motion. Walter Noll introduced the concept of motion of a continuum body B as a set of configurations $\{\theta_t\} \subset \Phi$, where t was a real number, called "time". The motion was a subject of two axioms:

N8X: The derivative $v(X, t) = \mathrm{d}\theta_t(X)/\mathrm{d}t$ exists for all $X \in B$ and all t, it is a continuous function of X and t jointly, and it is a smooth function of X.

N9X: The derivative $v'(X, t) = \mathrm{d}v(X, t)/\mathrm{d}t = \mathrm{d}^2\theta_t(X)/\mathrm{d}t^2$ exists piecewise and is piecewise continuous in X and t jointly [9.9].

Axioms of Force. The third group of axioms was postulated by Noll for the concept of force. The mathematical framework for his theory of forces was that of vector valued functions and measures. Noll differentiated between external (or body) forces and internal (or contact) forces. The concept of force comprised the concepts of body force and of contact force, and the value of this force was equal to the sum of the corresponding values of the body and contact forces. For each continuum body B, Noll introduced a system of *body forces* as a unity vector valued set of functions $\{f_b\}_{b\in B}$, satisfying the following axioms:

N10X: For each part b of B, f_b is a vector valued measure defined on the Borel subsets of b.

N11X: For each b, f_b is absolutely continuous relative to the mass distribution m of b. Hence it has a density s_b so that

$$f_b(G) = \int_G s_b(X)\mathrm{d}m \ .$$

N12X: The density s_b is bounded, i.e. $|s_b(x)| < k < \infty$, where k is independent of b and $X \in b$ [9.10].

Every continuum body B was a subject to actions of another system of forces – the *contact forces*, which were, according to Noll, represented as a unity of vector valued set functions $\{f_c\}_{c\in B}$, satisfying the following axioms (x^{cl} means the boundary of x):

N13X: For each part c of B, f_c is a vector valued measure defined on the Borel subsets of c.

N14X: $f_c(G) = f_c(G \cap c^{\mathrm{cl}})$.

N15X: If $d \subset c^{\mathrm{cl}}$, $d \subset e^{\mathrm{cl}}$, and $c \subset e$, then $f_c(d) = f_e(d)$.

N16X: If $\phi \in \Phi$ is any configuration of c^{cl} and if $P = \phi(c^{\mathrm{cl}})$, then the induced measure $f_c\phi^{-1}$, when restricted to the

Borel subsets of the boundary surface P^{cl} of $P = \phi(c)$, is absolutely continuous relative to the Lebésque surface measure on P^{cl}. Hence it has a density $s(c, \phi)$ so that

$$f_c(d) = \int_{\phi(d)} s(c, \phi; x)\mathrm{d}A$$

for all Borel subsets $d \subset c^{cl}$.

N17X: The density $s(c, \phi)$ is bounded, i.e. $|s(c, \phi; x)| < l < \infty$, where l does not depend on c or $x \in \phi(c^{cl})$ [9.11].

Noll formulated *three principles of continuum mechanics*. He defined a dynamical process as a triple $\langle B, \theta_t, F_{B,t} \rangle$, where B was a continuum body, θ_t stood for its motion, $F_{B,t}$ was a time-dependent system of forces for B.

Principle of Linear Momentum:
For all parts b of B and all times t

$$F_{B,t}(c) = \mathrm{d}/\mathrm{d}t\, g(c; t) = \mathrm{d}/\mathrm{d}t \int_c v(X, t)\mathrm{d}m$$

In words: The resultant force acting on the part c is equal to the rate of change of the linear momentum of c.

Principle of Angular Momentum:
Let $0 \in E$ be any point in space. Then for all parts c of B and all times

$$M(F_{B,t}, \theta_t, 0; c) = \mathrm{d}/\mathrm{d}t\, h(c; t; 0)$$
$$= \mathrm{d}/\mathrm{d}t \int_c [\theta_t(X) - 0]v(X, t)\mathrm{d}m$$

(M is the resultant moment about 0 acting on c) . . .
In words: The resultant moment about 0 acting on a part c is equal to the rate of change of the angular momentum of c relative to 0 [9.12].

Principle of Objectivity:
If a dynamical process is compatible with a constitutive assumption then all processes equivalent to it must also be compatible with this

constitutive assumption. In other words, constitutive assumptions must be invariant under changes of frame [9.13].

9.2 Second Axiomatic of Rational Continuum Mechanics

Before we approach Noll's most known work {2.9}, it is necessary to consider one other paper of his entitled "La Mécanique Classique Basée sur un Axiome d'Objectivité" {2.22}. In this Noll proposed some improvements for his "tentative" axiomatical system of continuum mechanics [9.14]. He wrote that the report {2.22} included a new axiomatization of continuum mechanics the main idea of which concerned the principles of linear and angular momentum as above: "Recently, I have observed that one can reach a more simple and more natural axiomatization by replacing the axioms of balance with another axiom of objectivity which requires the invariance of the work under change of arbitrary frame of reference. This axiom is compatible with the physical realities if one regards the forces of inertia as real forces which are the interactions between the bodies in our solar system and the totality of objects in the rest of the universe. The inertial frames of reference no longer enter into the general part of the new axiomatization. The law of inertia is regarded as a constitutive postulate. The priviledged frames are explicitly linked with the position of the rest of the universe" [9.15].

Axioms of Material Universe. First of all, Noll enlarged his axiomatic system to a set of bodies which he called "material universe". Obviously, he made in {2.22} the first attempt to propose a proper mathematical framework for the set of bodies, having in mind to approach the case of contact interactions later [9.16]. The material universe was a triple $\langle U, \Phi, m \rangle$, where U was a set of elements, called "material points", Φ was a set of mappings from U into E, a three-dimensional Euclidean point-space, m was a function, defined on U. The elements of Φ were called "configurations of the universe". The values of the function m were called "masses" of the corresponding sets of material points. Noll postulated the following axioms:

N1Y: Φ is a one-to-one mapping.

N2Y: Φ has to possess any suitable properties which supplied U with a structure of the continuously differentiable manifold, isomorphed to E [9.17].

N3Y: m is a positive finite measure on U, defined on the parts of U.

N4Y: A *body* is a closed subset of U, which is sufficiently regular and doesn't itself have strictly closed subsets of the same mass as the body [9.18].

N5Y: If $B_1 \in U$ and $B_2 \in U$ and $m(B_1 \cap B_2) = 0$, then B_1 and B_2 are "separated bodies".

N6Y: If $B \in U$, there exists $B^\circ \in U$, such that B° is separated from B and $m(B \cup B^\circ) = m(U)$ [9.19].

Axioms of Force. The second group of axioms was related to the concept of force. A system of forces was a vector valued function, defined on two bodies from U, separated from each other. Its value $f(A, B)$ was called the force exercised by the body A to the body B. The following axioms were postulated:

N7Y: For each $B \in U$ there exists a vector valued, completely additive function f_B, defined on the subsets of B:

$$f_B(A) = f(A, B^\circ), \quad A \subset B \ .$$

N8Y: For each $B \in U$ there exists a vector valued, completely additive function $f^*{}_B$, defined on the subsets of the body's exterior B°:

$$f^*{}_B(A) = f(B, A), \quad A \subset B^\circ \ .$$

N9Y: $f_B(B) = \int_B \mathrm{d}f_B = f(B, B^\circ)$ is called the "resultant force", exercised on B;

$\int_B g \cdot df_B$ is a Stieltjes integral of a continuous function g, defined on B.

N10Y: The system of forces is a combination of "contact forces" and "distant forces" [9.20].

Space-Time of Continuum Mechanics. Noll introduced the following concepts relating to the space-time structure of continuum mechanics:

(i)   he called "event" a pair (x, t), where x was a point in E, t was a real number;
(ii) x was called "position" of the event;
(iii) t was called "time" of the event;
(iv) all possible (x, t) were called "space-time".

Axioms of Change of Frame. The central place in the new Noll's axiomatical system belonged to the concept of change of frame. A *change of frame* was an automorphism of the space-time whose properties were subject of the following axioms:

N11Y: The change of frame preserves the time-lapse between every pair of events (x, t), (x', t'): $t' = t + a$, where a is a real number.

N12Y: The change of frame preserves the spatial distance between every pair of simultaneous events (x, t), (x', t'): $x' = c(t) + Q(t)x$, where x, x', $c(t)$ are vectors in respect to an arbitrary origin in E, $c(t)$ is a point-valued function of t, $Q(t)$ is a function of t with values, being orthogonal transformations [9.21].

Axioms of Dynamical Process. The fourth group of axioms was concerned with the concept of a *dynamical process*. A dynamical process was a pair $\langle \chi_t, f_t \rangle$, where χ_t was a set of configurations of U, f_t was a variety of systems of forces for U.

N13Y: The velocity of $X \in U$ is $v(X, t) = d\chi(X, t)/dt$, which exists and is sufficiently regular. The set χ_t or the function $\chi(X, t)$ is called "motion" of U. The spatial point $x = \chi(X, t)$ is called "position" of the material point X at time t.

N14Y: The *working* per unity of time of the forces, exercised on the body B, is

$$w(B, t) = \int_B v(X, t) \mathrm{d}f_{t,B} \, ,$$

where $f_{t,B}$ is a vector valued measure, associated with f_t (see **N7Y**).

N15Y: (Fundamental Axiom of Dynamical Process)

For a dynamical process, the working is objective, i.e. for every body $B \in U$ the working $w(B, t)$ (see **N14Y**) is invariant under an arbitrary change of frame (see **N11Y**, **N12Y**), expressed in the form: for $\langle \chi_t, f_t \rangle$ and $\langle \chi'_{t'}, f'_{t'} \rangle$

$$\chi'_{t'} = c(t) + Q(t)\chi_t \, ,$$
$$f'_{t'} = Q(t)f_t \quad [9.22] \, .$$

Noll derived from these axioms two fundamental laws of mechanics: (i) the **law of balance of actions**: the resultant force exercised on the whole body was zero; (ii) the **law of balance of moments**: the resultant moment of forces exercised on the whole body was zero [9.23].

He thought that the axioms **N1Y–N15Y** were still too general for continuum mechanics and pointed out how to proceed with its axiomatization: "The mathematical structure developed thus far is not sufficient for the description of any particular physical situation. It introduces only a general language which is very convenient in the majority of situations of classical mechanics. To define the particular nature of a mechanical situation it is necessary to introduce 'constitutive postulates'" [9.24].

Principles of Continuum Mechanics. Noll formulated two constitutive postulates and a principle of material objectivity of continuum mechanics in the most general form:

N16Y: "(Constitutive Postulate of the First Sort)

This specifies the action of the exterior S° of the system S on the body B in S. It is assumed that the forces $f(B, B^\circ) = f_S(B)$ acting on the body B in S by the rest of the universe

S° depend in a simple manner on the motion of B relative to a frame of reference defined in respect to S°. This frame of reference is called the 'priviledged frame'" [9.25].

N17Y: "(Constitutive Postulate of the Second Sort)
This specifies the mutual action of continuous bodies in the material system S. The general form of such a postulate is a functional relation between the restrictions of functions $\chi(X, t)$ and $f(B, C)$ to $X \in S$ and $B \subset S, C \subset S$. One calls such a relation a 'constitutive equation'" [9.26].

N18Y: "(Principle of Material Objectivity)
A constitutive postulate must be objective, i.e. invariant under a change of the frame of reference: ... if a constitutive postulate is satisfied for a dynamical process $\langle \chi_t, f_t \rangle$, the same constitutive postulate must be satisfied for all dynamical processes $\langle \chi'_{t'}, f'_{t'} \rangle$ equivalent to $\langle \chi_t, f_t \rangle$" [9.27].

9.3 Third Axiomatic of Rational Continuum Mechanics

However, these were not the two axiomatics, which brought Walter Noll great national and international fame. In 1958, his fundamental axiomatical theory of the mechanical behavior of continuous media {2.9} was published and caused revolutionary changes in the theory and practice of continuum mechanics. The set of important corollaries, which can be derived from his axiomatic of continuum mechanics and the mathematical framework of this theory were superior to those of Georg Hamel [9.28].

Noll's mathematical theory of the mechanical behavior of continuous media included two main parts: a *general theory of constitutive equations and his first theory of simple materials*.

Axioms of Constitutive Equations. In {2.9}, Noll clarified the *physical background* of his general axiomatical theory of constitutive equations: "(i) The stress at a particle X should depend only on the physical state of an arbitrary small neighborhood of X. The state of parts distant from X should have no direct influence on the stress at X. This condition is implicit in the concept of contact force; (ii) The physical state of a body at a time t should depend only on its past history, i.e., on what happened to it at times $\tau \leq t$, and not on its future, i.e. on what will happen to it at times $\tau > t$. This condition expresses the causality of natural processes" [9.29]. He introduced the concept of the *physical history of a mechanical body*: "The physical history of a body consists of several components: its kinematical history, its thermodynamic history, its electromagnetic history, its chemical hisory, etc. In reality, each of these components may influence the stress" [9.30].

CE1: (Deformation)
(i) A smooth homeomorphism d which maps a neighborhood of the null-vector 0 of the vector space V onto another such neighborhood and which maps 0 into itself is called a local homeomorphism of V.
(ii) An equivalence relation among all local homeomorphisms is defined by local identity: $d \sim d'$ if and only if $d(P) = d'(P)$ for all P in some neighborhood of 0, however small. The resulting equivalence classes W will be called deformations" [9.31].

CE2: (Gradient of Deformation)
The gradient ∇ of a deformation W is defined by the formula

$$\nabla W = \nabla d(0), \quad \forall d \in W \quad [9.32] .$$

CE3: (Principle of Determinism)
The stress $S(t)$ at a particle X and at time t is determined by the past history of the motion of an arbitrary small neighborhood of X.
(i) The stress $S(t)$ at X is a functional F_t of the motion θ, the domain F_t being the class of all possible motions and its values being symmetric tensors;
(ii) for any two motions θ and θ' which coincide in some

neigborhood of X for times $\tau \leq t$ the value of the functional F_t is the same, i.e. $F_t(\theta) = F_t(\theta')$ whenever $\theta(Z, \tau) = \theta'(Z, \tau)$, $\tau \leq t$, Z is in some neighborhood $N(X)$ of X, however small [9.33].

CE4: (General Form of Constitutive Equation)
The most general form of constitutive equation is $S(t) = F(E^t)$, where E^t is the kinematical history of particle X for the local motion E, and the following condition is satisfied:

$$q_0 F(E^*) q_0^T = F(q^* \circ E^*), \quad q_0 = q^*(0)$$

for all kinematical histories $E^* \in K_{X^*}$ and all orthogonal tensor functions $q^* \in Q^*$. F is called the *functional of the particle X* [9.34].

CE5: (Isomorphism of Bodies)
An *isomorphism* of a body B onto a body B' is a one-to-one mapping γ of B onto B' such that
(i) the configurations ϕ' of B' are of the form $\phi' = \phi \circ \gamma^{-1}$;
(ii) the mass distributions m and m' of B and B' are related so that $m(c) = m'(\gamma(c))$ for all Borel subsets c in B [9.35].

CE6: (Material Isomorphism)
A local isomorphism Γ of a particle X onto a particle X' will be called a *material isomorphism* of X onto X' provided the functionals F and F' of X and X' are related by

$$F(E^*) = F'(E^* \circ \Gamma^{-1})$$

for all kinematic histories $E^* \in K_{X^*}$ [9.36].

CE7: Two particles consist of the same material if they are materially isomorphic to each other [9.37].

CE8: (Constitutive Functional)
A functional G whose domain is the set D^* of all deformation histories and whose values are symmetric tensors is called a *constitutive functional* provided it has the following property:

For all deformation histories $\Omega^* \in D^*$ and all orthogonal tensor functions $q^* \in Q^*$ the relation

$$q_0 G(\Omega^*) q_0{}^T = G(q^* \circ \Omega^*), \quad q_0 = q^*(0)$$

holds [9.38].

CE9:   (Local Isotropy Group)
Let G be a constitutive functional. The group A^1 of all isochoric deformations $W \in T$ with the property that $G(\Omega^* \circ W) = G(\Omega^*)$ holds for all $\Omega^* \in D^*$ is called the *local isotropy group* of G [9.39].

Axioms of Simple Materials. Noll's first theory of simple materials included the following axioms:

SM1:   A constitutive functional G is said to be simple if $G(\Omega^*) = G(\nabla\Omega^*)$ for all deformation histories $\Omega^* \in D^*$ [9.40].

SM2:   The material at a particle X is *simple* or, briefly, X is simple if the constitutive functional G_Φ of X, for some local configuration Φ of X, is simple [9.41].

SM3:   The group A of all unimodular transformations H with the property that $G(F^* H) = G(F^*)$ holds for all $F^* \in L^*$ is called the *isotropy group* of the simple constitutive functional G [9.42].

SM4:   (Simple Solid)
It is said that a constitutive functional G defines a *solid* if its isotropy group is a subgroup of the orthogonal group, i.e. if $A \subset Q$ [9.43].

SM5:   (Solid Particle)
A particle X is said to be a *solid particle* if there is a configuration gradient M of X such that G_M defines a solid [9.44].

SM6:   (Isotropic Solid)
(i) A solid is called an *isotropic solid* if the isotropy group of its defining functional is the full orthogonal group, i.e. if $A = Q$.

(ii) A solid is called *anisotropic* if the isotropy group of its defining functional is a proper subgroup of the orthogonal group, i.e., if $A \subset Q$, $A \neq Q$.

(iii) Any configuration gradient M such that $G_M = Q$ is called an *undistorted state* of X [9.45].

SM7: (Simple Fluid)
It is said that a constitutive functional defines a *fluid* if its isotropy group A is the full unimodular group U, i.e., if $A = U$ [9.46].

SM8: (Fluid Particle)
A particle X is said to be a *fluid particle* if, for some configuration gradient M of X, the corresponding constitutive functional defines a fluid, i.e. if $A_M = U$ [9.47].

As Noll showed in {2.9}, the concept of simple material generalized most of the standard types of materials, for example, materials of differential type [9.48].

9.4 Walter Noll, Mario Bunge and a New Theory of Simple Materials

After the success of Walter Noll's first axiomatics of continuum mechanics, Mario Bunge, whom Clifford Truesdell described once as "the only philosopher he knew who had any contact with real science", proposed his axiomatical system for continuum mechanics [9.49]. It may be assumed that Mario Bunge obtained this set of axioms through the axiomatization technique which included sixteen essential steps [9.50]. As merits of Noll's axiomatics of continuum mechanics of 1957/59, Bunge pointed out their mathematical modernity, rigor and origin from the functional analysis. However, as he wrote, Noll didn't care much for physical contents or, better, if he did it, then he "borrowed uncritically the physicist's operationalist doctrine of meaning, according to which the testability was necessary for meaningfulness rather than the other way around" [9.51].

The purpose of Bunge's axiomatization attempt was to make the physical content of continuum mechanics more apparent in comparison with Noll [9.52]. Bunge included in his axiomatical system nine basic primitives, twenty-one other concepts of continuum mechanics, and three groups of axioms: the axioms of body (seven axioms), the kinematic axioms (three axioms) and the axioms of dynamics (four axioms). It was "more schematic and mathematically less sophisticated" than that of Noll [9.53]. Since Bunge "borrowed freely" from the works of Noll for his axiomatic, one can easily conclude that the delicate defects of Noll's axiomatics of continuum mechanics remained a deep secret for him [9.54]. Bunge's axiomatic made only a very slight impact on continuum mechanics. His idea to rearrange the axioms of continuum mechanics to the form in which they looked as a result of application of his axiomatization technique proved itself to be impotent.

However, Walter Noll himself estimated the success of his first axiomatics as a moderate one, though, as he pointed out, it "served as a foundation for a large part of the research in continuum physics since 1958". Its conceptual tools and notations began from that time to be used "routinely and without reference in textbooks on continuum mechanics" [9.55]. It was not Bunge's axiomatical system which moved Noll to elaborate a new mathematical theory of simple materials. With the exception of a short period between 1964 and 1965, Bunge and his works seemed to remain out of the view and interests of Noll. The great mathematician had no respect for the philosophy of science, and his knowledge of philosophy didn't go beyond the standard university program. Noll's first axiomatics of continuum mechanics possessed, as it was stated in the introduction to {2.37}, "at least three severe defects". The main defect was "that the present stress was determined by the infinite past history of deformation". However, in the comments on it Noll invoked only "philosophical" grounds for the non-existence of the infinite memory [9.56]. Two other aspects were discovered by D.R. Owen ("inadequate conceptual framework for the mathematical description of ... plasticity, yield, and hysteresis") and by B. Bernstein (Noll's framework as artificial and fraught with difficulties in defining a material of the rate type), who proposed theories without these defects. Though Noll accepted in principle these critical remarks, he refused to recognize these theories as adequate solutions of the problems. He explained that Bernstein's theory simply didn't fit into the framework of his first axiomatics of continuum mechanics. He also found Owen's solution to be complicated and vague, though it didn't contradict the first theory of simple materials and described "what physicists usually called plastic behavior" [9.57]. The purposes of the *new theory of simple materials* were set by Noll as: (i) to consider deformation processes only of finite duration; (ii) to use the primitive notion of a "state of the material element", characteristic for most branches of physics; (iii) to make the description of material elements invariant in respect to the space-time structure, which could be classical, neo-classical or relativistic; (iv) to make the theory of simple materials free from a priori frames of reference, coordinate systems and the physical space [9.58]. It was published in 1972 as {2.37} and began with a comprehensive section containing numerous preliminary mathematical notations, theories of vector spaces, bilinear forms, orthogonal groups and inner-product spaces [9.59]. Perhaps it is because of this complicated mathematical framework, which demands from its user an ability to think mathematically, that Noll's new theory of simple materials hasn't yet found wide applications in continuum mechanics.

9.5 Walter Noll's Second Theory of Simple Materials

SM1N: (Process)
The process is a function from a closed segment $[0, a]$ into an arbitrary set; the process has a duration of a, initial and final values; when the initial and final values coincided for every

$$[0, t] \subseteq [0, a] ,$$

such a process is called "freeze of duration a"; the processes can be divided into segments and be continued.

SM2N: (Body element)

A body element is a triple (A, B, C), where A is a finite-dimensional real vector space, B is a closed and connected subset of the set of all positive definite symmetric bilinear forms on A and C is a class of processes with values in B which satisfy the following conditions:

(i) any freeze at any $b \in B$ belongs to C;

(ii) if $c \in C$, so does every segment of c;

(iii) C is closed under continuation;

(iv) any two elements $b_1, b_2 \in B$ can be connected by a process in C; the elements of B are called *configurations* of a body element A and every element of C the *deformation process* for A [9.60].

SM3N: (Placement)

A placement of the material (A, B, C) is called an element D of the set of invertible linear mappings from A to a real inner product space V of the same dimension as A, called the frame space, if $D^*D \in B$, where D^* is a dual element of D.

SM4N: (Motion)

A motion of the material element (A, B, C) in V is a process

$$c : [0, a] \rightarrow E_V$$

with the property that $c^*c \in C$ [9.61].

SM5N: (Intrinsic stress)

The intrinsic stress S is an element of the space of symmetric bilinear forms on A, such that

$$S = D^{-1}LD^{*-1} ,$$

where L is a linear mapping from V to V, such that $L = L^{\mathrm{T}}$, where L^{T} is the transpose of L [9.62].

SM6N: (Material element)

A material element is a septuple (A, B, C, W, X, Y, Z) which satisfies six axioms; the nature of the objects of the septuple is:

(i) (A, B, C) is a body element; it is called the *underlying body element* of the material element;

(ii) W is a set, called the *state space* of the material element;

(iii) X is a mapping from W into B;

(iv) Y is a mapping from W to the space of symmetric bilinear forms on A, and this space is called the *stress space* of the material element;

(v) Z is a mapping from the set of all state-process pairs such that the state "fitted" the initial configuration of the process to W [9.63].

SM6.1N: The state reached after a process should fit the final configuration [9.64].

SM6.2N: When the material element is first subjected to a process P_1 and then to a process P_2 it should reach the same state as when it was subjected to the continuation $P_1 * P_2$ of P_1 by P_2 [9.65].

SM6.3N: There must be some operational way to distinguish between states; specifically, if two states are different but fit the same configuration, there must be some process which produces different stresses with the two states as initial states [9.66].

In order to expand his axiomatical theory so as to include the case of non-zero internal constraints, Walter Noll formulated a normalization condition, which removed the indeterminacy of the intrinsic extra stress, a natural substitution to the intrinsic stresses Y in the definition of material elements [9.67]. With the help of a "general notion of isomorphism for arbitrary mathematical structures", he introduced definitions of isomorphisms among the body and the material elements. They allowed him to present a relation of "consisting of the same material" and to clarify the term "material" in the cartesian spirit: "If two material elements ... are materially isomorphic we also say that they *consist of the same material*. We give a precise meaning to the term 'material' by saying that a *material* is an equivalence class of material elements, the equivalence being material isomorphy" [9.68]. If for every two material elements of a material body they consisted of the same material, then the body should be called, according to Noll, *materially uniform* [9.69]. Noll transferred the theories of material symmetry and isotropy to his new axiomatics and developed the necessary terminology [9.70]. With the help of some general topological works, he constructed the conceptual tools to deal with the uniformity and topology in the state space. However, his attempt wasn't very successful. The price he was to pay for it was a technical fourth axiom of the material element, which didn't have any physical sense [9.71].

SM7N: (Relaxed state)
          If the material element is initially in a certain state and then frozen in its configuration, its state will approach, in the limit of infinite time, a relaxed state [9.72].

SM8N: Every state must be accessible from some relaxed starting state [9.73].

In order to prove the efficiency of the new theory of simple materials, Noll considered in the paper {2.37} its application to a vast number of special cases, including basic theories of semi-elastic and elastic materials, differentiation of solid and fluid states, deformation histories, monotonous processes (including deformation processes of constant reduced state and state of monotonous flow), material functionals for incompressible fluids and materials of the rate type [9.74].

One of the most important parts in {2.37} was its Section 21, which Walter Noll devoted to a discussion of "possible generalizations, further development, and applications of the theory" [9.75]. They included such fields as: (i) *theories of fading memory*, which could be built on the basis of the new axiomatic of simple materials with the help of the principle of analogy from the classical theories of fading memory; as Noll remarked, these reformulations were in no way trivial; as an example here, he pointed to the theories of fading memory for semi- and non-semi-elastic materials; (ii) *theories of plastic behavior*, where he proposed to describe their conceptual tools in terms of the new theory of simple materials; he expected the results to be more natural, simple, and clear in comparison with those of D.R. Owen; (iii) *thermodynamic theories of material elements*, for which he gave precise recommendations on how to build them as generalizations of the new theory of simple materials; he expected such theories "to simplify and to clarify the axiomatic approach to thermodynamics developed by W.A. Day"; (iv) *abstract theories of systems with memory*: Noll formulated a proposal for a general concept of a *physical system with memory*, which he differentiated from the structure of J.C. Williams' mathematical systems theory [9.76].

10 Rational Fluid Mechanics

The success of Noll's axiomatics of continuum mechanics depended largely upon its efficiency in applications to standard mechanical problems. Between 1959 and 1966, Noll was working intensively in the field of fluid mechanics. The overhelming majority of his publications on that subject was prepared in collaboration with Bernard D. Coleman, now one of the prominent scientists of modern American rational mechanics. Coleman was five years younger than Noll. Before their first meeting in 1957, Coleman had worked at Yale University and then as a research chemist at the Carothers Research Laboratory of the Du Pont Company. Having been educated as a physical chemist, he was interested first in the kinetics of chemical substances (at Yale University) and later in polymer mechanics (at the Mellon Institute). However, Coleman's knowledge of mathematics was at that time very slight [10.1].

The first paper of Noll and Coleman on fluid mechanics was finished at the beginning of March 1959 and was devoted to certain steady flows of general fluids [10.2]. The purpose of the paper was to show that steady flow problems, such as simple shearing flow, channel flow, Poiseuille flow and Couette flow, could be solved assuming only Noll's definition of a general fluid from {2.9} and the incompressibility [10.3]. Noll and Coleman imposed some restrictions on motions and material properties: (i) motions were to be isochoric; (ii) the stress in fluid didn't depend on mass density in the explicit form; (iii) fluids were to be homogeneous and incompressible; (iv) body forces per unit mass were to possess a single-valued potential [10.4]. Under these restrictions they obtained explicit expressions for the physical components of the stress tensor in the fluid through three certain material functions τ, σ_1, σ_2, which depended on the fluid material itself, but not on a particular

type of flow. As Noll and Coleman showed, the material functions
satisfied the following requirements: (i) τ was to be an odd function
and to have the same sign as χ, a second order tensor, specified
for the steady flows; if τ was assumed to be twice continuously
differentiable then $\tau(\chi)$ was to be a strictly increasing function of
χ in a segment $[-\chi_0; \chi_0]$; in the same segment, τ was to possess
a strictly increasing and odd inverse function τ^{-1}; $\tau(0) = 0$; (ii)
$\sigma_1(\chi)$ and $\sigma_2(\chi)$ were to be even functions, such that

$$\sigma_1(0) = \sigma_2(0) = 0 \quad [10.5].$$

Noll and Coleman expressed a rheological function, called "shear-
dependent viscosity", in terms of the three material functions and
could then clarify its general meaning [10.6].

They obtained dynamical equations for the rectilinear flow of
the fluid and considered two practically valuable special cases of
it: simple shearing flow and flow through a channel [10.7]. The first
case corresponded to the motion of adherent fluid between an infinite
plate at rest and another infinite plate moving parallel to it with a
constant velocity. Under a physically plausible hypothesis on the
pressure distribution inside the fluid, Noll and Coleman derived a
formula showing that the velocity of flow was a linear function and
didn't depend on time. Their formula for the shearing stress allowed
them to determine the material function τ experimentally. They also
obtained a set of equations which "at least in principle" could be
used to evaluate the other two material functions σ_1 and σ_2 from the
measurements of normal stresses [10.8]. The flow through a channel
was assumed to be a rectilinear flow of adherent fluid between two
parallel infinite plates at rest. Using the general expressions for the
rectilinear flow, Noll and Coleman deduced a formula for velocity
profile. They estimated the volume discharge per unit time through a
cross-section of the channel of one unit depth as an integral formula.
Further analysis led them to the conclusion that the experimentally
measurable rate of discharge was related to the measurable driving
force a (i.e. the applied force per unit volume in the direction of the
flow) through the material function τ, which could be determined
through an independent measurement. Noll and Coleman derived a
formula for calculation of τ, when the volume discharge had been

found as a function of a, and also some formulas for normal stresses during motion [10.9].

The two other cases in the paper were the Poiseuille flow and the Couette flow. The Poiseuille flow was a steady flow of adherent fluid through an infinite circular pipe of a given radius. At the first step, Noll and Coleman obtained the dynamical equations for the problem and, after some elementary mathematical reasoning, the conditions of their compatibility. Under the condition that the shearing stress was continuous at the center of the pipe, they derived some formulas for the velocity profile in an arbitrary section of the pipe and for the volume discharge per unit time through a cross-section of the pipe, depending on the driving force a and the material function τ. Noll and Coleman specified an expression which allowed them to calculate τ if the volume discharge could be measured. They also gave formulas for normal stresses during the motion of the fluid [10.10]. The Couette flow was defined as a steady flow of fluid between two infinite coaxial cylinders ($r_1 \leq R \leq r_2$). Noll and Coleman obtained the dynamical equations of the process and also certain formulas for the moment per unit height exerted on the fluid inside a cylindrical surface $R =$ const, and for the derivative of the velocity of fluid particles. For a special case, when the outer cylinder (r_2) rotated with a constant angular velocity and the inner cylinder (r_1) remained at rest, they received: (i) an integral formula for the calculation of the angular velocity as a function of τ and of the moment per unit height; (ii) an approximate formula for the velocity without sign of integral under the condition that ($r_2 - r_1$) was sufficiently small; (iii) two formulas for the calculation of τ in the Couette flow from the experimental data; (iv) formulas for normal stresses in the fluid; (v) an integral and an integral-free formula for the difference of normal stresses on the surfaces of the outer and inner cylinders in the case when ($r_2 - r_1$) was sufficiently small [10.11].

The second paper of Noll and Coleman on fluid mechanics was published in October 1959. It was devoted to the helical flow of Noll's general fluids [10.12]. Noll and Coleman formulated the concept of helical flow in the introduction to the article: "Helical flow is a steady laminar flow which can occur in an annular mass of fluid which is contained between two infinite coaxial cylinders of (different) radii ... which rotate around their common axis with (certain) angular velocities ... , while a pressure gradient or external body force is acting on the fluid parallel to the axis of rotation" [10.13]. The helical flow included as special cases the concentric-pipe flow under zero applied torque in the axial direction, the Couette flow with a zero driving force, applied in the axial direction, simple shearing flow for the infinite radii of the cylinders and for the zero axial driving force, and channel flow under infinite radii of the cylinders and for zero torque [10.14]. The main purpose of the paper was to show that the three material functions τ, σ_1, σ_2 from {2.11} determined completely the stress and velocity profile for the helical flow of adherent fluids [10.15].

It is worth mentioning here that Noll and Coleman achieved re-markable progress in comparison with the internationally recognized studies of R.S. Rivlin on helical flows in 1955–1956. They wrote proudly: "Rivlin presented his results in terms of eight material functions. We showed that only three were necessary" [10.16]. The helical flows of Noll and Coleman met several requirements. To some degree they repeated the restrictions imposed in the earlier paper {2.11}. The new demands concerned the properties of extra stress which Noll introduced in his Ph.D. thesis in 1954 [10.17]. Noll and Coleman described the physical sense of all three material functions in terms of the rheological concepts: σ_1 and σ_2 determined the "normal stress effects" and τ was related to the shear-dependent viscosity [10.18]. They deduced explicit mathematical formulas for the physical components of extra stress in terms of these material functions. Using the expressions of these components as functions of radius only, Noll and Coleman transformed the dynamical equations of helical flow and obtained the compatibility conditions for the equations of the transformed system. They derived a formula for the moment per unit height exerted on the fluid inside the cylindrical surface of constant radius. With the help of this formula the scientists managed to clarify the physical background of the moment: it was the torque per unit height, required to maintain the relative motion of the bounding cylinders. Noll and Coleman calculated a formula for the driving force and made clear its physical sense, when the total applied force in the axial direction was exerted on the annulus of the volume of the fluid between the two planes, perpendicular to the axial direction. If the body forces vanished, the driving force was a pressure head per unit length in the axial direction. If the gravity was the only applied axial force, the driving force was equal to the specific weight of the fluid. The scientists derived formulas for the complete velocity profile as a function of the radius and for the volume discharge per unit time through a cross-section, perpendicular to the axial direction. Noll and Coleman also derived a formula connecting the difference of the angular velocities of the inner and outer cylinders to the torque per unit height and to the axial driving force. They added to it a mathematical expression for the physical components of the stress tensor in the fluid. In this study, Noll and Coleman specified in an explicit form the relationships among their material functions τ, σ_1, σ_2 and the eight material functions used by R.S. Rivlin in his study of the helical flows. At the same time, they established correspondences among τ, σ_1, σ_2 and the four material functions proposed for helical flow by H. Markovitz, a colleague of Coleman at the Mellon Institute [10.19].

In 1961, a summary {2.16} by Noll and Coleman was published which included their "results in the mechanics of fluids that could be of interest to polymer physicists" [10.20]. The scope of the paper included phenomenological aspects of the mechanical behavior of viscoelastic (not-obeying the classical laws of fluid mechanics or elasticity) fluids under the assumption that "all non-mechanical influences could be neglected" [10.21]. Noll and Coleman made substantial comment on the concept of a simple fluid. Noll even managed to include in it a new form of definition of a simple fluid, equivalent to that of his first axiomatization of continuum mechanics [10.22]. The partners concentrated their attention on incompressible fluids, whose motions always happened with a volume-preservation. They reviewed first the case of simple shearing flow. Noll and Coleman introduced the concept of simple shearing flow, the matrix of physical components of extra stress, the material functions τ, σ_1, σ_2 and their properties. Some special forms of those functions were discussed for different types of incompressible fluids: perfect, Newtonian, Reiner-Rivlin fluids and fluids of differential type. The scientists demonstrated how to use material functions for the solution

of an elementary boundary value problem. Then they turned to other
types of flows: the Poiseuille flow, flow between concentric pipes
and the Couette flow. The exact formulas were given for velocity
profile, volume discharge per unit time through a cross-section and
for the physical components of the stress tensor in terms of the three
material functions. Noll and Coleman also pointed out the known
close connection of their theories with experimental fluid mechanics. Noll and Coleman turned to the analysis of a class of periodic
shearing flows and established that, in this case, the extra stress was
to be a periodic function with the same period as the flow itself.

On the basis of their early theory of fluid motions, Noll and Coleman formulated some fundamental ideas on the role of the theory
of Newtonian fluids in rheology: (i) Newtonian fluids approximated
the behavior of nearly all real fluids in the limit of slow motions; (ii)
the *logical status of the theory of Newtonian fluids*: "when the kinematical history didn't differ too much from the 'rest history' ... ,
the theory·of Newtonian fluids should be a first-order correction to
the theory of perfect fluids" [10.23]. In {2.16} one can find an important remark of Noll and Coleman on the experimental value of
their theory of fading memory: "All real materials have an imperfect
memory: in any experiment we expect the events that happened in
the recent past to be more important than those that happened in the
very distant past. Indeed if this were not the case it would be exceed-
ingly difficult to design experiments, because the experimenter, not
being in control of events that happened before he was born, would
never be sure that he had adequate knowledge of the history of the
materials he was testing" [10.24]. Noll and Coleman formulated a
mathematical "fundamental smoothness assumption" on the constitutive functionals *H* occurring in the definition of a simple fluid and
derived some corollaries from it [10.25]. The review {2.16} ended
with a thorough analysis of the higher-order corrections to the theory of perfect fluids [10.26].

In 1962, the paper {2.19} by Noll and Coleman on the steady
extension flows of incompressible simple fluids was published. The
term "steady extension" had the following meaning: "We say that
the fluid body under consideration is undergoing *steady extension*
if there exists a fixed Cartesian coordinate system such that the

components v_i of the velocity of the material point at x_1, x_2, x_3 are given by equations of the form $v_i = a_i x_i$, where the a_i are constants obeying the relationship $a_1 + a_2 + a_3 = 0$ [10.27]. As it follows from the definition, the steady extension flow was a new one in the mechanics of fluids and possessed special properties: (i) it was isochoric; (ii) it wasn't a viscometric flow or a simple shearing flow; (iii) it represented an example of substantially stagnant motion for which the general theory of simple fluids didn't reduce itself to the theory of a particular Rivlin-Ericksen fluid [10.28]. Noll and Coleman evaluated the components of the stress tensor for this type of flow in terms of two material functions, different from τ, σ_1, σ_2, known to us. They established some properties of the new material functions: (i) for a general incompressible simple fluid, they weren't determined in terms of τ, σ_1, σ_2; (ii) for a Newtonian fluid, one of the material functions vanished, and another one was equal to the doubled value of viscosity of the fluid [10.29].

The scientists managed to derive a necessary and sufficient condition that the power of dissipation for steady extension was non-negative in terms of the material functions. They obtained explicit formulas for the components of the stress tensor in a steady extension flow and formulated an important recommendation for rheologists: "In order to calculate the tractions which must be applied to the surface of a body to sustain a steady extension we must specify the body force potential and the shape of the body" [10.30].

Noll and Coleman also considered two applications: extension of a box and of a circular cylinder. The first problem was related to a fluid body which "had a shape of a box whose edges were parallel to the coordinate axes", and the lengths of the edges of the box, parallel to x_1 axis were functions of time. Noll and Coleman derived formulas for normal stress (per unit area of the present configuration) differences, and they indicated how these formulas could be used to calculate the two material functions for a steady extension flow from the experimental data [10.31]. The second problem corresponded to the case of a fluid body inside a curvilinear cylinder with the axis of rotation coinciding with the x_1 axis. The radius and the length L of the cylinder were taken to be definite functions of time. Noll and

Coleman evaluated the normal stresses at the two bounding cross sections and on the cylindrical boundary and came to the following conclusion: "One cannot maintain steady extension without applying tractions normal to the cylindrical surface. ... if these tractions are not supplied the cylinder will tend to bulge for $0 \leq x_1 \leq L/\sqrt{3}$ and to contract for $L/\sqrt{3} < x_1 \leq L$" [10.32].

In April 1962, Coleman presented a joint report with Noll called "Simple Fluids with Fading Memory" {2.26}, at the International Symposium on Second-Order Effects in Elasticity, Plasticity and Fluid Dynamics in Haifa, Israel. It was a summary of "some aspects of the mathematical theory of fluid behavior they had developed during" 1959–1962, and in particular of their theory of constitutive equations in the case of fluid bodies [10.33]. This report is a valuable source of definitions of the basic concepts of the theory of simple fluids in ordinary language: (i) "A *simple material* is a substance for which the present stress is determined by the history of the strain; (ii) a *fluid* is a substance with the property that all local states with the same mass density are intrinsically equivalent in response, with all observable differences in response being due to definite differences in history; (iii) the *principle of fading memory*: deformations which occurred in the distant past should have less effect on the present value of the stress than deformations which occured in the recent past; (iv) a material obeying (i) and (ii) is a *simple fluid*; (v) a material obeying (i), (ii) and (iii) is a *simple fluid with fading memory*" [10.34].

Noll and Coleman formulated in ordinary language the general requirements of simple fluids with and without fading memory: (i) "*basic constitutive assumption of classical hydrostatics*: the stress on a fluid which has remained at rest at all times is a hydrostatic pressure depending only on the density; (ii) *graduate stress relaxation*: consider a fluid which until time 0 has been subjected to various deformations (which we need not specify) and suppose that for all times t greater than 0 the fluid is in a fixed configuration; then, as $t \to \infty$ the stress in the fluid gradually decays to a hydrostatic pressure which depends only on the density; (iii) *Newtonian behavior in slow motions*: in the limit of very slow flows the stress in a fluid is approximately given by the constitutive equation of a linearly viscous fluid; (iv) *linear viscoelastic behavior in infinitesimal deformations*: if the history of a fluid happens to be such that all the configurations the fluid experiences are infinitesimally close to each other, then the stress in the fluid is approximately given by Boltzmann's ... classical theory of infinitesimal viscoelasticity (with the equilibrium shear modulus set equal to zero)" [10.35].

The report consisted of two parts. The first included the mathematically exact expositions of the basic concepts of the theory of simple fluids with fading memory: deformation gradient, relative deformation gradient, relative right Cauchy-Green tensor, reduced history of stress tensor, stretching tensor, simple fluid, simple fluid always at rest, fading memory, simple fluid with fading memory, statical extension, stress relaxation and others [10.36]. The second part of the report was concerned with applications. Noll and Coleman introduced first the concepts of perfect and Newtonian fluids, coefficients of viscosity, retardation of a reduced history, and retardation factor. They formulated a theorem on Newtonian behavior where the retardation factor played the central role. They turned after that to the case of finite linear viscoelasticity and infinitesimal viscoelasticity of fluids. They wrote out a theorem on infinitesimal viscoelastic behavior and ended the report with a general analysis of finite second-order viscoelasticity for fluids [10.37].

It would appear the collaboration between Noll and Coleman on fluid mechanics approached its end in 1961. The best proof of this fact is their separate publications on one and the same type of fluid flow. Coleman called it "substantially stagnant motion", while Noll preferred the term "motion with constant stretch history" as more distinct and related to his theory of simple fluids. There can be no doubt for all those who know the art of collaboration of Noll with other scientists that the idea and basic theory of such flows were discussed between Noll and Coleman before these publications. It should be assumed that they had originally planned to join their results in a comprehensive theory. However, Coleman was a researcher in Noll's sense of the term (see Chapter 6); he was not afraid of premature publications when the issue of priority was at stake. Noll probably considered his own results and those of Coleman as being insufficient for a more or less complete theory in

1961. Noll wanted to wait and, in fact, allowed Coleman to enjoy
the laurels of priority.

Noll's paper on fluid motions with constant stretch history appeared in 1962 [10.38]. He began it with a conceptual analysis which
also included the central concept of "motion with constant stretch
history": "We say that a *motion*, along the path of a material point,
has *constant stretch history* if there is an orthogonal tensor function
$P(t)$ such that the histories $C_t^t(s)$ and $C_o^o(s)$, for $s \geq 0$, are related
by

$$C_t^t(s) = P^T(t)C_o^o(s)P(t), \quad s \geq 0.$$

The condition ... means physically that for an observer moving with
the material point both the magnitudes of the principal stretches and
the changes of direction of the principal axes of strain depend only
on the time lapse s and not on the present time t" [10.39]. The main
result of the paper was a representation theorem for the deformation
gradient for motions with constant stretch history: "A motion has
constant stretch history if and only if the deformation gradient $F_o(\tau)$
relative to some fixed time 0 has the representation

$$F_o(\tau) = Q(\tau)\exp(\tau M), \quad Q(O) = I,$$

where M is a constant tensor and $Q(\tau)$ is an orthogonal tensor function" [10.40]. Noll gave four equivalent mathematical definitions of
a fluid motion which he called "locally viscometric". It was in this
paper that the concept of viscometric flow appeared. Noll wrote: "A
viscometric flow is a flow which is locally viscometric at every material point of the flowing medium" [10.41]. He introduced the concept of steady curvilinear flow and proved that such a flow was a
special type of viscometric flow [10.42]. He contributed to the general theory of simple fluids and proved a result on the characterization of the behavior of an incompressible simple fluid for motions
with constant stretch history [10.43].

In 1963, Noll gave an invited lecture at the First International
Symposium on Pulsatile Blood Flow. He reviewed in it the classical and modern theories of flow in tubes which could be useful in
the biomechanical studies of blood circulation. Noll divided his report into three parts, according to the three classes of fluids which

he considered: perfect fluids, linearly viscous fluids, and simple fluids. He described the research history of motion of a perfect fluid in a tube. For the linear viscous fluids he wrote out the expressions for the stress tensor and pointed out their correspondence with the famous Navier-Stokes equations. With their help he reviewed some central results related to the simple shearing and Poiseuille flows. In the last section of the lecture, devoted to simple fluids, Noll listed some properties of the stress tensor and several key results relating to the cases of simple shearing and Poiseuille flows. In the discussion which followed the lecture, Noll described the status of his theory of fluid motions: "This is an extremely general theory, and it starts from almost no assumptions at all. Almost anything can happen. The variation of the radial pressure depends on what the viscometric functions are, and these functions, of course, depend on the fluid. In the classical theory the behavior of the fluid in all conceivable situations – at least in mechanics – is determined by one constant alone, the shear viscosity. That is all you have to know, and in principle it is then only a matter of solving problems, so to speak, to obtain the behavior of the fluid for all possible flows. The situation is very different for the general theory. One constant is not enough here; you need much more information. For instance, for shearing flow, Poiseuille flow, and several other flows, you need three functions. These three functions must be determined experimentally and then tabulated. ... Several people have measured them for various fluids, usually for substances that are suitable for precise measurements. Polyisobutylene solution is such a substance, and Markovitz at the Mellon Institute is carrying out an extensive program of measurement for the viscometric functions of polyisobutylene solution. Even the three viscometric functions are insufficient to describe the behavior of the fluid in flows other than steady viscometric flows. You can have two fluids that behave the same way in steady viscometric flows but behave differently in other experiments, for example, in stress relaxation experiments" [10.44].

In 1966, a monograph {1.3} by Noll, Coleman and Markovitz on the viscometric flows of non-Newtonian (simple) fluids was published by the Springer-Verlag. It contained a completely new method of Noll's to deal "with viscometric flows without the apparatus of relative Cauchy-Green tensors and reduced constitutive equations" [10.45]. Noll and his collaborators wrote on the scope of the monograph: "This is a book about viscometric flows of simple fluids. We endeavor to give here a complete and self-contained presentation of the existing theory of viscometric flows and to discuss in detail experimental techniques for realizing them in the laboratory" [10.46].

The book received much acclamation. It was called fundamental, remarkable, pleasant, outstanding for the first penetration into modern non-linear continuum mechanics. A.C. Pipkin, a famous specialist in rational mechanics, wrote about it in 1966: "Modern engineering practice has made it necessary to understand and deal with a wide variety of materials whose behavior is sufficiently complex that the classical constitutive equations of linear elasticity or Navier-Stokes fluid dynamics are not useful descriptions. The diverse possibilities of non-linear behavior make it doubtful that any comparatively simple constitutive equation can do more than describe limited aspects of the properties of materials. For this reason, some investigators are turning to the analysis of specific classes of motions for which a complete description of all material properties is neither necessary nor even desirable. The monograph at hand represents an outstandingly successful example of this approach. The presentation here is a short but complete, detailed, and self-contained description of both the theoretical and experimental aspects of the subject, written by leading contributors to it. Specific, experimentally determined forms of the viscometric functions for a variety of materials are provided as illustrations. There is an extensive bibliography and a short history of the experimental and theoretical development of the subject. The book will presumably become the standard reference on the subject" [10.47]. The main theoretical achievement of the book was Noll's analysis of the new general concept of fluid mechanics: the viscometric flow [10.48]. He showed that the theory which had been created earlier for the steady, helical and other flows, could be expanded to the case of viscometric flows. The majority of the theoretical sections of the monograph was a review of the known publications of Noll and Coleman on fluid mechanics between 1959 and 1966 [10.49].

11 Rational Elasticity

The first of Walter Noll's works on rational viscoelasticity and finite elasticity, written in collaboration with B.D. Coleman, was published in April 1961. It was devoted to a reexamination of "fundamental hypotheses of linear viscoelasticity in the light of recent advances in non-linear continuum mechanics" [11.1]. This paper {2.17} included, under the influence of their earlier article {2.15}, special axioms for the *concept of fading memory*:

VE1: (Postulate of Fading Memory)
There exists an influence function $h(s)$ of an order $r > 0.5$ such that, for each value of the tensor parameter c, the functional F of the constitutive equation of the simple material is Fréchet-differentiable at the zero history in the Hilbert space $\mathfrak{H}$ corresponding to $h(s)$ [11.2].

VE2: The Fréchet-differentiability of F postulated in the **VE1** is uniform in the tensor parameter c [11.3].

VE3: The tensor function $h(c)$ in the constitutive equation of the simple material is continuously differentiable [11.4].

Noll and Coleman derived from the general theory of simple materials and the postulates **VE1–VE3** the *constitutive equation of infinitesimal viscoelasticity* which generalized the classical equation:

$$T(t) - T_r = W(t)T_r - T_r W(t) + [\Omega + \Phi(0)]\{E(t)\}$$
$$+ \int_0^\infty (\mathrm{d}/\mathrm{d}s)[\Phi(s)]\{E(t-s)\}\mathrm{d}s \ ,$$

where $T(t)$ was the stress tensor, $T_r = h(I)$ was the residual stress (i.e., the stress the material would sustain if it had been held in the reference configuration at all times in the past), $W(t) = 0.5(H -$

H^T), $H = F - I$, $F = F(\tau)$ was the deformation gradient, Ω was the gradient of the tensor function $h(c)$ at $c = I$, $E(\tau)$ was the infinitesimal strain tensor and

$$E = 0.5(H + H^T) , \quad \Phi(s) = -2 \int_s^\infty \Gamma(\sigma)\mathrm{d}\sigma ,$$
$$(\mathrm{d}/\mathrm{d}s)[\Phi(s)] = 2\Gamma(s) ,$$

where $\Gamma(s; c)\{\ldots\}$, for each s and each c, was a linear transformation of the space of symmetric tensors into itself with the property

$$\int_0^\infty \mathrm{tr}[\Gamma(s; c)\Gamma^T(s; c)]h(s)^{-2}\mathrm{d}s < \infty \quad [11.5] .$$

The rational *finite linear viscoelasticity* was a theory based on the constitutive equation of a simple material in the form

$$T = h(c) + \int_0^\infty \Gamma(s; c)\{G(s)\}\mathrm{d}s ,$$

where $G(s)$ was a history (see axiom **FM2** in Chapter 12) [11.6]. It allowed Noll and Coleman to compare the infinitesimal and the finite linear theories of viscoelasticity. They wrote: "Aside from the fact that the finite theory ... applies to a much larger class of problems than the infinitesimal theory, there is a fundamental difference between the two theories. The infinitesimal theory is physically meaningless for finite deformations because it does not have the invariance properties required by the principle of material objectivity. The finite linear theory, on the other hand, enjoys the correct invariance. Thus, it is conceivable that there exists *some* material which obeys equation (for finite linear viscoelasticity) ... for *arbitrary* finite deformations. The infinitesimal theory cannot possibly apply to *any* material when finite deformations are considered" [11.7]. For the case of isotropic materials, they proved the following "natural" (see {2.15}) fact: "In the finite theory of linear viscoelasticity, the behavior of an isotropic material is determined by eleven independent scalar material functions; three of these depend on three variables and the remaining eight on four variables" [11.8].

Noll and Coleman obtained the *constitutive equation of a simple fluid in the theory of finite linear viscoelasticity*:

$$T = -p(\rho)I + \int_0^\infty \mu(s, \rho)J(s)\mathrm{d}s + \left[\int_0^\infty \lambda(s, \rho)\mathrm{tr}\, J(s)\mathrm{d}s\right]I ,$$

where $p(\rho)$, $\mu(s, \rho)$, $\lambda(s, \rho)$ were the three scalar material functions determining the mechanical behavior of a fluid. They considered its special case to be when the fluid was assumed to be incompressible and established a correspondence between the relaxation function, known from the rheology, and $\mu(s)$ [11.9].

They analysed in detail the case in which the axiom **VE1** was enlarged to the 2-times Fréchet differentiability of the functional F. If this new axiom was postulated, then Noll and Coleman derived the *constitutive equation of a simple fluid in the second-order theory of viscoelasticity* and its special case to be when the fluid was incompressible. The stress tensor of incompressible fluid was defined again in terms of three scalar material functions. Noll and Coleman managed to establish a connection between their theory of fluid mechanics and second-order viscoelasticity.

The Noll-Coleman theory led to an important insight into the field of different viscoelasticity theories: (i) "Different choices of the measure of strain (C_t or C_t^{-1} or $\log C_t$ etc.) corresponded to different theories of finite linear viscoelasticity; however, the difference of the stresses computed using two different such theories was of order $o(\|G(s)\|)$; (ii) Different choices of the measure of strain ... corresponded to different theories of second-order viscoelasticity; these different theories were equivalent in the sense that the corresponding stresses differ only in terms of order $o(\|G(s)^2\|)$" [11.10].

Noll and Coleman pointed out the differences between their theory of viscoelasticity and that of R.S. Rivlin and his collaborators [11.11]. The second-order (and also higher-order) theories of viscoelasticity obtained in {2.17} were used in the next joint paper by Noll and Coleman {2.18} to make "predictions concerning the normal stresses derived from the second-order theory of incompressible viscoelastic fluids" [11.12].

In 1965, Walter Noll made a report "The Equations of Finite Elasticity" {2.28} at a Symposium on Applications of Non-Linear

Partial Differential Equations in Mathematical Physics, held by the American Mathematical Society. It represented a summary of "a comprehensive modern exposition of the theory of finite elastic deformations" in his joint treatise with C.A. Truesdell on the non-linear field theories of mechanics {1.2}. The following axioms were included by Noll in his conceptualization of finite elasticity:

FE1:   (i) The material at a material point X in a body is said to be *elastic* if for all possible dynamical processes the present stress T at X is determined by the present local configuration M at X: $T = t(M)$;

(ii) If we choose a fixed reference configuration κ, the relation $T = t(M)$ is equivalent to the *stress relation* $T_R = h(F)$, where F is the present deformation gradient at X and T_R the present Piola-Kirchhoff stress at X; h is the *response function* of the material at X with respect to κ [11.13].

FE2:   A *body* is said to be *elastic* if the material at *all* of its points X is elastic [11.14].

FE3:   The *isotropy group* G of an elastic material consists of all unimodular tensors H such that

$$h(FH)H^{\mathrm{T}} = h(F)$$

holds identically for F in the domain of h [11.15].

FE4:   An *elastic fluid* is defined by the property that its isotropy group coincides with the full unimodular group: $G = U$ [11.16].

FE5:   The response function h has a potential, i.e. there exists a scalar-valued stored-energy function $\sigma(F)$ such that

$$h(F) = \nabla\sigma(F) ,$$

where ∇ denotes the gradient operator in the nine-dimensional space L of all tensors. Such a material is called *hyperelastic* [11.17].

Noll derived from Cauchy's law of motion a *differential equation for a homogeneous elastic body* and gave its form in several useful standard cases. For different types of elastic bodies he proposed four boundary value problems for these equations:

(i) "If we assume that the traction field t_R is independent of S ... , we obtain a *boundary condition of fixed traction*; in this case, the surface traction per unit area in the reference configuration is held constant in magnitude and direction no matter how the body may deform;

(ii) In a *boundary condition of pressure* it is assumed that the traction acting on a point of S is normal to S and depends, let's say, on the position of the point or on the volume of the region bounded by S;

(iii) *Boundary condition of surface action*: Let Σ be the set of all surfaces that can be obtained by deforming the boundary ∂B from its reference configuration; let A be the mapping which assigns to every $s \in \Sigma$ a vector field $t_R = A(s)$, whose values $t_R(X, t) = T_R(X, t)n_R$ give the surface traction to be applied should the boundary ∂B assume the configuration s at time t;

(iv) *Boundary condition of place*: This is always a prescription of the configuration of the boundary surface ∂B: $\chi(X, t) = \chi_b(X, t)$, where $\chi_b(X, t)$ is a given function defined for $X \in \kappa(\partial B)$ and all $t \geq 0$" [11.18].

Noll stressed that he "wasn't aware of any investigations concerning the existence or uniqueness" of the differential equation for homogeneous elastic bodies under the specified boundary conditions (i)–(iv) and the initial conditions. He wrote: "It is often reasonable to expect *local* uniqueness, i.e., it is reasonable to expect that in a certain neighborhood of a given solution there is no other solution satisfying the same boundary conditions and coinciding with the given solution at some point" [11.19]. In {2.28}, he gave a description of the basic ideas on how to reduce statical boundary value problems to variational problems in the case of hyperelastic bodies. He stressed that he "wasn't aware of any work presenting significant *sufficient* conditions for the existence of stable solutions", i.e. of solutions to the minimization problem [11.20].

12 Rational Thermomechanics

All Noll's papers between 1959 and 1966 relating to the field of thermomechanics (thermodynamics of materials), were written in collaboration with B.D. Coleman [12.1]. The latter remembered in 1990: "In the early 1950's, for graduate courses in physical chemistry at Yale, I read the then standard texts on thermodynamics and selections from the works of Gibbs. I recall being awed by the generality of the subject; it appeared to have deep implications in every branch of macroscopic physics; but I also recall being uneasy about the completeness or clarity (I was not sure which) of its principles" [12.2].

Their first joint paper was already completed in Febrary 1959 and was published in the July. It was entitled "Conditions for Equilibrium at Negative Absolute Temperatures" {2.13}. Its main result was a modification of the so-called "energy theorem". In its common form, it asserted that "the state of equilibrium of a system with a given entropy was that in which the system had its *lowest* energy". In the general case, which also included negative absolute temperatures, Noll and Coleman showed that "the equilibrium states had *minimum* or *maximum* energy among all states with equal entropy" [12.3].

As the next step in an axiomatization of thermomechanics, Noll and Coleman developed "a rigorous theory of thermostatics for continuous bodies in arbitrary states of strain" [12.4]. Their theory was to be differentiated, on the basis of its physical principles, from those of J.W. Gibbs, J. Hadamard, J.L. Ericksen, R.A. Toupin, R. Hill and P. Duhem. Noll and Coleman introduced the following axioms:

TS1:   (Global Thermomechanic State)
A global thermomechanic state, or simply a *state*, of a body B is a pair $\{f, \eta\}$, consisting of a configuration f of B and a scalar field η defined on B; η is called the *entropy distribution* of the state [12.5].

TS2:   (Local Thermomechanic State)
A local thermomechanic state, or simply a *local state*, of a material point X is a pair (M, η) consisting of a local configuration M of X and a real number η, called the *entropy density* (per unit mass) of the local state [12.6].

TS3:   The local configuration transforms under a change of frame according to the law $F' = QF$ where Q is orthogonal [12.7].

TS4:   The entropy density η remains invariant under a change of frame [12.8].

TS5:   (Equivalence of States)
(i) Two local states are equivalent if they only differ by a change of frame of reference;
(ii) The global states are equivalent if they only differ by a change of reference [12.9].

TS6:   A material is characterized by a real valued function of local states, whose values ε are called the *energy densities* (per unit mass) of the local states: $\varepsilon = \varepsilon^*(F, \eta)$, where $F = MM_r^{-1}$, where M stands for a local configuration of the body, M_r is its reference configuration [12.10].

TS7:   The energy density is invariant under a change of frame [12.11].

TS8:   (Caloric Equation of State)
$\varepsilon = \varepsilon^*(F, \eta) = \varepsilon^*(U, \eta)$, where U is the right stretch tensor [12.12].

TS9:   The energy function e^* remains the same function if the local reference configuration M_r is changed to another local reference configuration $M_r' = HM_r$ with the same

density:
$$\varepsilon^*(F, h) = \varepsilon^*(FH, \eta) \,,$$

where H is an unimodular transformation [12.13].

TS10: The unimodular transformations H from **TS9** form a group, called the *isotropy group* G of ε^* or of the material defined by ε^* [12.14].

TS11: (Simple Fluid)
(i) One says that the energy function ε^* defines a *simple fluid* if its isotropy group G is the full unimodular group U;
(ii) A material point is called a *fluid material point* if its energy function defines a simple fluid;
(iii) The caloric equation of fluid state:

$$\varepsilon^* = \varepsilon(v, \eta) \,, \quad v = 1/\rho = |\det F|(1/\rho) \,,$$

where v is called the "specific volume" of the local configuration $M = F M_r$, ρ and ρ_r are the mass densities in M and M_r [12.15].

TS12: (Simple Solid)
(i) It is said that the energy function ε^* defines a *simple solid* if its isotropy group G is contained as a subgroup in the orthogonal group O;
(ii) A material point is called a *solid material point* if its energy function ε^*, relative to some local configuration as a reference, defines a simple solid [12.16].

TS13: (Mechanical Equilibrium)
(i) In order that a body B be in *mechanical equilibrium* under a given system of forces, two conditions must be fulfilled for each part P of B:
(a) (Force Condition)
the sum of the forces acting on P must vanish; this condition depends on the body and the force system, it does not depend on the body's configuration;
(b) (Moment Condition)

the sum of the moments, about any point, of the forces acting on P must vanish; this condition depends on the body's configuration;

(ii) It is said that a material point X is in *local mechanical equilibrium* when the body is in a given configuration and under a given force system if the stress tensor s exists at X and is symmetric [12.17].

TS14:   (Thermal Equilibrium)

The local state (F, η) is called a state of thermal equilibrium under a given force temperature pair (s_r, θ), where s_r is the Kirchhoff tensor of a system of contact forces at X, θ is a real number to be interpreted as the temperature at X if:

(i) the stress tensor $s = (\rho/\rho_r)Fs_r$ is symmetric;

(ii) the inequality $\lambda^*(F^*, \eta^*) > \lambda^*(F, \eta)$ holds for all states $(F^*, \eta^*) \neq (F, \eta)$ such that $F^* = GF$ where G is symmetric and positive definite;

$$\lambda(F, \eta) = \varepsilon^*(F, \eta) - (1/\rho_r)\text{tr}(Fs_r) - \eta\theta \ ,$$

i.e., $\lambda(F, \eta)$ is the energy density at a local state minus the potential energy, per unit mass, of the local contact forces minus "thermal potential energy" [12.18].

TS14′:   (Thermal Equilibrium)

The local state (F, η) is a state of thermal equilibrium under the force temperature pair (s_r, θ) iff the following three conditions are satisfied:

(i) The stress tensor $s = (\rho/\rho_r)Fs_r$ is given by the *stress relation*

$$s = \rho F \varepsilon^*{}_F(F, \eta) \ ;$$

(ii) The temperature θ is given by the *temperature relation*

$$\theta = \varepsilon^*{}_\eta(F, \eta) \ ;$$

(iii) The inequality

$$\varepsilon^*(F^*, \eta^*) - \varepsilon^*(F, \eta) - \text{tr}[(F^* - F)\varepsilon^*{}_F(F, \eta)]$$
$$-(\eta^* - \eta)\varepsilon^*{}_\eta(F, \eta) > 0$$

holds if $(F^*, \eta^*) \neq (F, \eta)$ and F^* is related to F by $F^* = GF$, where G is positive definite and symmetric [12.19].

TS14″: (Thermal Equilibrium)
The local state (F, ε) is called a state of thermal equilibrium under the force temperature pair (s_r, θ), with $\theta \neq 0$, if:
(i) the stress tensor $s = (\rho/\rho_r)Fs_r$ is symmetric;
(ii) the inequality

$$\eta^{**}(F, \varepsilon) > \eta^{**}(F_1, \varepsilon_1) + (\theta\rho_r)^{-1}\text{tr}[(F_1 - F)s_r]$$
$$-(\varepsilon_1 - \varepsilon)/\theta$$

holds for all states $(F_1, \varepsilon_1) \neq (F, \varepsilon)$ such that $F_1 = GF$, where G is symmetric and positive definite, is a solution of the caloric equation of state [12.20].

TS15: (First Fundamental Postulate of Thermostatics of Continua)
(i) For every local state (F, η) for which $\varepsilon^*(F, \eta)$ is defined there exists a force temperature pair (s_r, θ) such that (F, η) is a state of thermal equilibrium under (s_r, θ).
(ii) (Physical Meaning)
At a material point, any local thermomechanic state can be an equilibrium state provided the local temperature and local forces have appropriate values [12.21].

TS16: (Second Fundamental Postulate of Thermostatics of Continua)
(i) The energy function $\varepsilon^*(F, \eta)$ is strictly increasing in η for each fixed F.
(ii) (Physical Meaning)
At least in continuum mechanics, absolute temperatures are never negative [12.22].

Noll and Coleman considered applications of their conceptualization of thermostatics of materials to the studies of infinitesimal deformations from an arbitrary state, simple fluids, isotropic materials and free energy [12.23]. They called the *main purposes of thermostatics*: (i) the exploration of the consequences for the caloric equation of state of the existence of local states of thermal equilibrium; (ii) the derivation of useful, necessary and sufficient criteria for global states to be stable [12.24]. In order to approach the second purpose, they were to add to their axiomatic some appropriate postulates, clarifying *thermal and mechanical stability*. It was to be differentiated between the concepts of *equilibrium and stability*, which had often been confused in scientific literature [12.25].

TS17:   (Thermal Stability)
(i) Let $\{f, \eta\}$ be a state of B and let E and H be, respectively, the total internal energy and total entropy corresponding to the state $\{f, \eta\}$. We say that $\{f, \eta\}$ is a *thermally stable* state of B if every other state $\{f, \eta^*\}$, with the same configuration as $\{f, \eta\}$ and the same total entropy as $\{f, \eta\}$,

$$H^* = \int_B \eta^*(X)\mathrm{d}m = H = \int_B \eta(X)\mathrm{d}m \ ,$$

has a greater total internal energy than the state $\{f, \eta\}$;

$$\text{i.e., } E^* = \int_B \varepsilon(X, \eta^*(X))\mathrm{d}m > E = \int_B \varepsilon(X, \eta(X))\mathrm{d}m \ .$$

(ii) (Equivalent Definition)
A state $\{f, \eta\}$ of B is called *thermally stable* iff every other state $\{f, \eta^*\}$ with the same configuration as $\{f, \eta\}$ and the same total energy as $\{f, \eta\}$,

$$E^* = \int_B \varepsilon(X, \eta^*(X))\mathrm{d}m = E = \int_B \varepsilon(X, \eta(X))\mathrm{d}m$$

has a lower total entropy than the state $\{f, \eta\}$; i.e.,

$$H = \int_B \eta(X)\mathrm{d}m > H^* = \int_B \eta^*(X)\mathrm{d}m \quad [12.26] \ .$$

TS18:   (Isothermal Stability at Fixed Boundary or IFB Stability)
An equilibrium state $\{f, \eta\}$ is called *IFB stable* if $\{f, \eta\}$

has a uniform temperature θ and if for every state $\{f^*, \eta\}$ which satisfies the following conditions:

(i) f^* lies in a prescribed neighborhood of f, defined by the metric

$$\delta(f, f^*) = \sup_{X \in B} \{|f^*(X) - f(X)| + |F^{*-1}(X)F(X) - I|\}$$

over the space of all configurations;

(ii) $f^*(X) = f(X)$, when X belongs to B^*;

(iii) the temperature corresponding to $\{f^*, \eta\}$ is equal to θ for all X in B; the following inequality holds:

$$\int_B \{\Psi(F^*) - \Psi(F) - b(f^* - f)\}dm \geq 0 .$$

Here B^* is the boundary of B and $\Psi(F) = \Psi^*(F(X), \theta; X)$; Ψ^* is the free-energy function defined by

$$\Psi^*(F, \theta) = \varepsilon^*(F, \eta^*(F, \theta)) - \theta \eta^*(F, \theta) ;$$

$F^*(X)$ and $F(X)$ are the deformation gradients at X for the configurations f^* and f, respectively, both computed relative to the same fixed reference configuration [12.27].

TS19: (Isothermal Stability at Fixed Surface Tractions or IFT Stability)

An equilibrium state $\{f, \eta\}$ is called *IFT stable* if $\{f, \eta\}$ has a uniform temperature θ and if for every state $\{f^*, \eta\}$ which satisfies the following conditions:

(i) f^* lies in a prescribed neighborhood of f;

(ii) the temperature corresponding to $\{f^*, \eta\}$ is equal to θ for all X in B, the following inequality holds

$$\Gamma = \int_B \{\Psi(F^*) - \Psi(F) - b(f^* - f)\}dm$$
$$- \int_{B^*} (f^* - f)s\mathbf{n}dA \geq 0$$

Here dA is the element of the surface B^*, and $\mathbf{n}$ is the exterior unit normal [12.28].

TS20:   (Adiabatic Stability at Fixed Boundary or AFB Stability)

1) An equilibrium state $\{f, \eta\}$ is called *AFB stable* if $\{f, \eta\}$ is thermally stable and if for every state $\{f^*, \eta^*\}$ which satisfies the following conditions:

(i) f^* lies in a prescribed neighborhood of f;

(ii) $f^*(X) = f(X)$, when X belongs to B^*;

(iii) $\int_B \eta^*(X)\mathrm{d}m = \int_B \eta(X)\mathrm{d}m$;

the following inequality holds:

$$\int_B \{\varepsilon^*[F^*(X), \eta^*(X); X] - \varepsilon^*[F(X), \eta(X); X]$$
$$- b(f^* - f)\}\mathrm{d}m \geq 0. \quad (\circ)$$

$\{f, \eta\}$ is *strictly AFB stable* iff $(\circ)$ is strict for all $\{f^*, \eta^*\}$ satisfying (i)–(iii) and for which $f^* \neq f$.

2) (Alternative Definition)

A thermally stable equilibrium state $\{f, \eta\}$ is *AFB stable* iff for every state $\{f^*, \eta^*\}$ which satisfies the following conditions:

(i) f^* lies in a prescribed neighborhood of f;

(ii) $f^*(X) = f(X)$ when X belongs to B^*;

(iii) $\int_B \{\varepsilon^*(F^*(X), \eta^*(X); X) - bf^*\}\mathrm{d}m$
$\quad\quad = \int_{BX} \{\varepsilon^*(F(X), \eta(X); X) - bf\}\mathrm{d}m$;

the following inequality holds:

$$\int_B \eta^*(X) \cdot \mathrm{d}m \leq \int_B \eta(X) \cdot \mathrm{d}m . \quad (*)$$

Furthermore, $\{f, \eta\}$ is *strictly AFB stable* iff $(*)$ is a strict inequality for every state $\{f^*, \eta^*\} \neq \{f, \eta\}$ obeying (i)–(iii) [12.29].

TS21:   (Adiabatic Stability at Fixed Surface Tractions or AFT Stability)

1) An equilibrium state $\{f, \eta\}$ is called *AFT stable* if it is thermally stable and if for every state $\{f^*, \eta^*\}$ which satisfies the following conditions:

(i) f^* is in a prescribed neighborhood of f;

(ii) $\int_B \eta^*(X) dm = \int_B \eta(X) dm$;
the following inequality holds:

$$\int_B \{\varepsilon^*[F^*(X), \eta^*(X); X] - \varepsilon^*[F(X), \eta(X); X]$$

$$- b(f^* - f)\} dm - \int_{B^*} (f^* - f) s\boldsymbol{n} dA \geq 0 .$$

2) If the inequality in 1) holds for all states which obey (i) and (ii) and is a strict inequality for all such states for which $F^*(X) \neq F(X)$ for at least one X, then $\{f, \eta\}$ is *strictly AFT stable against deformations and rotations* [12.30].

Finally, Noll and Coleman defined "a type of stability which was proposed by Gibbs for fluids free from body forces" and obtained two types of necessary and sufficient conditions for an equilibrium state to be stable in the J.W. Gibbs sense [12.31].

The history of their next major paper {2.15} on continuum mechanics was presented by Coleman in 1985 [12.32]. This work is considered here since the historical roots of the theory of fading memory lie in the attempts of Noll and Coleman to formulate a plausible thermodynamics of simple materials [12.33]. The conceptual system of the **theory of fading memory**, presented in {2.15}, included the following postulates:

FM1: (Influence Function)
An *influence function h* of order $r > 0$ is a real-valued function of a real variable with the following properties:
(i) $h(s)$ is defined and continuous for $0 \leq s < \infty$;
(ii) $h(s)$ is positive, $h(s) > 0$;
(iii) for each $\sigma > 0$, there is a constant M_σ, independent of α, such that $\sup_{s \geq \sigma} \{h(s/\alpha)/[\alpha^r h(s)]\} \leq M_\sigma$ for $0 < \alpha \leq 1$
[12.34].

FM2: (History)
A *history* is a measurable function $g = g(s)$ defined for $0 \leq s < \infty$ with values in s, a real Banach space [12.35].

FM3:   (Retardation)

The *retardation* Γ_α with retardation factor α, $0 < \alpha \leq 1$, is the linear transformation $g \to g_\alpha$ defined, for all histories g, by

$$(\Gamma_\alpha g)(s) = g_\alpha(s) = g(\alpha s) \quad [12.36] .$$

FM4:   (Generalized Derivatives)

Generalized derivatives of g of the order $k = 0, 1, \ldots, n$ are the limits

$$\overset{(0)}{g} = \lim_{s \to 0} g(s) ;$$

$$\overset{(k)}{g} = \lim_{s \to 0} (k!/s^k)[g(s) - \sum_{j=0}^{k-1} (s^j/j!) \overset{(j)}{g}] \quad [12.37] .$$

FM5:   (Taylor Transformation)

The *Taylor transformation* Π_n is the linear transformation $g \to \Pi_n g$ defined for all histories g which are n times differentiable at $s = 0$, by

$$(\Pi_n g)(s) = \sum_{j=0}^{n} (s_j/j!) \overset{(j)}{g}$$

where the $g^{(j)}$ are the generalized derivatives [12.38].

FM6:   (Memory Functional)

A function F defined on a neighborhood in H of the zero function $0 \in H \subset L_{h,p}$ and having values in a real Banach space T will be called a *memory functional of type* (h, n) if it is n times Fréchet-differentiable at $0 \in H$ and if it is normalized by

$$\delta^\circ F(g) = F(0) = 0 .$$

H is the function space of all histories g with the following properties:

a) g has a finite $L_{h,p}$-norm;
b) g has n generalized derivatives at $s = 0$;
c) g has a zero limit at $s = 0$: $\lim\limits_{s \to 0} g(s) = \overset{(0)}{g} = 0$;
d) n, p and the order r of the influence function h obeys the inequality:
$n < r - (1/p)$ and $(1/p) = 0$ if $p = \infty$.

$L_{h,p}$ is the Banach space of all histories with finite $L_{h,p}$-norms:

$$\|g\|_{h,p} = \sup_{s \geq 0} |g(s)| h(s) \,, \quad p = \infty \,;$$

$$\|g\|_{h,p} = \left\{ \int_0^\infty (|g(s)| h(s))^p ds \right\}^{1/p} \,, \quad 1 \leq p < \infty \,.$$

(iii) $\delta^\circ F(g) = F(0)$.

$$\delta^k F(g) = k! \lim_{\lambda \to 0} (1/\lambda^k)[F(\lambda g) - \sum_{j=0}^{k-1} (\lambda_j/j!) \delta^j F(g)]$$

are the Fréchet-differentials of F at $0 \in H$ [12.39].

One of the important results of the paper {2.15} was the so-called approximation theorem which "permitted the asymptotic approximation of a memory *functional*, for 'slow' histories, by a polynomial *function* of the derivatives at $s = 0$ of the argument function of the functional" [12.40]. The paper included a new requirement to be added to the definition of a simple fluid in Noll's first axiomatization of continuum mechanics. Noll and Coleman wrote: "We require that the functionals occuring in this definition be memory functionals in the sense of the definition used in this paper" [12.41]. The *constitutive equation for simple fluids* was written out in its explicit form:

$$S(t) = -p(\rho(t))I + \mathop{\mathfrak{F}}_{s=0}^{\infty} (c_t(t - s) - I, \rho(t)) \,,$$

where I stood for the unit tensor, $p(\rho)$ was a scalar function, $S(t)$ was the stress, $\rho(t)$ was the density at time t, $c_t(\tau)$ was the right Cauchy-Green tensor at time τ relative to the configuration at time t,

$$\mathop{\mathfrak{F}}_0^{\infty}(0; \rho) = 0 \quad [12.42] \,.$$

Noll and Coleman managed to show that the corrected theory of simple fluids led to a general result: "For simple fluids a finite number of scalar material functions sufficed to determine the stress to within terms of order n in α" [12.43]. They also obtained the *constitutive equation of an isotropic simple material with fading memory* in the form

$$S(t) = h(B(t)) + \mathop{\mathfrak{I}}_{s=0}^{\infty} (c_t(t-s) - I; B(t)) \,,$$

where $B(t)$ was the left Cauchy-Green tensor, taken relative to an undistorted reference state, and gave recommendations on how to obtain the constitutive equation of an anisotropic simple material [12.44].

In February of 1963, Noll and Coleman finished a joint paper on the thermodynamics of continuous media {2.21}. Coleman remembered: "It was not until the Fall of 1962 that Noll and I found, after several false starts, the methods presented in the paper {2.21} ... for deriving conditions necessary and sufficient for a given class of constitutive assumptions to be compatible with the second law. In {2.21} we applied the method to a class of materials for which (in contrast to materials with gradually fading memory) the implications of the second law were already considered self-evident" [12.45]. This paper contained a "physical description of the thermodynamics of materials". Noll and Coleman wrote in it: "To discuss the *thermodynamics* of continua, it appears that to the concepts of continuum mechanics one must add five new basic concepts: these are *temperature*, specific *internal energy* (sometimes called 'internal energy density'), specific *entropy* (sometimes called 'entropy density'), *heat flux*, and *heat supply* (sometimes called 'density of absorbed radiation') .... Once mechanics is axiomatized, it is easy to give the mathematical entities representing the thermodynamic concepts: temperature, specific internal energy, specific entropy and heat supply are scalar fields defined over the body, while heat flux is a vector field over the body. We believe that in presenting thermodynamics one should retain all the general principles of mechanics but add to them two new principles: the first law of thermodynamics, i.e. the law of balance of energy, and the second law, which for continua takes the form of the Clausius-Duhem inequality. The constitutive assumptions of the present work are the following: (i) That there exists a caloric equation of state relating the specific internal energy to the 'strain' (more precisely, the deformation gradient) and the specific entropy; (ii) That the stress tensor of mechanics is the sum of two terms, one of which, the 'elastic' term, depends on only the strain and the specific entropy, and the other, the 'viscous' term, depends on both these variables and, in addition, has a linear dependence on the rate of 'strain' (more precisely, the velocity gradient); (iii) That there exists a temperature equation relating the temperature to the strain and specific entropy; (iv) That the heat flux depends on only the strain, the specific entropy and the spatial gradient of the temperature. (We assume smoothness but not linearity for this dependence.) We allow the heat supply to be assignable in any way compatible with the general principles, just as body forces are often left assignable in mechanics" [12.46]. The following axioms were postulated for the *thermodynamics of simple materials*:

TD1: (Spatial Position)
         The spatial *position* $x = \chi(X, t)$, X is a material point, t – the time; the function χ is called the *deformation function*; it describes a *motion* of the body [12.47].

TD2: The symmetric stress tensor $T = T(X, t)$ [12.48].

TD3: (Body Force)
         The *body force* $b = b(X, t)$ per unit mass (exerted on the body by the external world) [12.49].

TD4: (Specific Internal Energy)
         The specific *internal energy* $\varepsilon = \varepsilon(X, t)$ [12.50].

TD5: (Heat Flux)
         The *heat flux* vector $q = q(X, t)$ [12.51].

TD6: (Heat Supply)
         The *heat supply* $r = r(X, t)$ per unit mass and unit time (absorbed by the material and furnished by radiation from the external world) [12.52].

TD7:   (Specific Entropy)
The *specific entropy* $\eta = \eta(X, t)$ [12.53].

TD8:   (Local Temperature)
The local temperature $\theta = \theta(X, t)$, which is assumed to
be always positive: $\theta > 0$ [12.54].

TD9:   (Thermodynamic Process)
The set of eight functions in **TD1–TD8** defines a *thermodynamic process* if the following equations are satisfied
for each part of the body B:
(i) **The law of balance of linear momentum**:

$$\int_B x'' \mathrm{d}m = \int_B b\,\mathrm{d}m + \int_{\partial B} T\boldsymbol{n}\mathrm{d}s \; ;$$

(ii) **The law of balance of energy**:

$$0.5(\mathrm{d}/\mathrm{d}t)\int_B x'x'\mathrm{d}m + \int_B \varepsilon'\mathrm{d}m$$
$$= \int_B (x'b + r)\mathrm{d}m + \int_{\partial B}(x'T\boldsymbol{n} - q\boldsymbol{n})\mathrm{d}s \; ,$$

where ∂B is the surface of B, $\mathrm{d}s$ is the element of surface
area in the configuration at time t, $\boldsymbol{n}$ – the exterior unit
normal vector to ∂B in the configuration at time t [12.55].

TD10:   (Admissible Thermodynamic Process)
A thermodynamic process is said to be *admissible* in a
homogeneous body consisting of an elastic material with
heat conduction and viscosity if the following constitutive
equations hold at each material point X and at all times t:

$$\varepsilon = \varepsilon^*(F, \eta) \; ,$$
$$\theta = \theta^*(F, \eta) \; ,$$
$$T = T^*(F, \eta) + l(F, \eta)[L] \; ,$$
$$q = q^*(F, \eta, \mathrm{grad}\theta) \; ,$$

where F denotes a deformation gradient at X and t, computed relative to a fixed homogeneous reference configuration; L is the velocity gradient, i.e. if one identifies X

with its position X^* in the reference configuration, he has

$$F = \nabla_{X^*} \chi(X^*, t) \quad \text{and} \quad L = F' F^{-1} \ ;$$

the value $l(F, \eta)$ of the response function l is a linear transformation over the nine-dimensional space of tensors, and the square brackets indicate that this transformation operates on the tensor L [12.56].

TD11: (Elastic Material with Heat Conduction and Viscosity)
An *elastic material with heat conduction and viscosity* is defined by five *response functions* $\varepsilon^*, \theta^*, T^*, l, q^*$ which depend on the choice of the reference configuration [12.57].

TD12: (Isotropy Group)
The *isotropy group* T of an elastic material with heat conduction and viscosity is the set of all unimodular tensors H for which the following identities hold:

$$\varepsilon^*(F, \eta) = \varepsilon^*(FH, \eta) \ ,$$
$$\theta^*(F, \eta) = \theta^*(FH, \eta) \ ,$$
$$T^*(F, \eta) = T^*(FH, \eta) \ ,$$
$$l(F, \eta)[L] = l(FH, \eta)[L] \ ,$$
$$q^*(F, \eta, v) = q^*(FH, \eta, v) \ ,$$

for all scalars η, all vectors v and all tensors F and L [12.58].

TD13: (Clausius-Duhem Inequality)
For every process admissible in a body consisting of a given material and for every part B of this body the inequality $\Gamma \geq 0$ is valid, where Γ is the production of entropy and

$$\Gamma = (\mathrm{d}/\mathrm{d}t) \int_B \eta \,\mathrm{d}m - \int_B (1/\theta) r \,\mathrm{d}m + \int_{\partial B} (1/\theta) q \boldsymbol{n} \,\mathrm{d}s \ ,$$

or, under suitable smoothness assumptions,

$$\Gamma = (\mathrm{d}/\mathrm{d}t) \int_B \gamma \,\mathrm{d}m \ ,$$

$\gamma = \eta' - (r/\theta) + (1/\rho)\text{div}(q/\theta) = \eta' - (r/\theta)$
$+ [1/(\theta\rho)]\text{div}q - -[1/(\rho\theta^2)]q\,\text{grad}\theta$ – (the specific production of entropy),
(q/θ) is a vectorial *flux of entropy*, (r/θ) is a scalar *supply of entropy* [12.59].

Noll and Coleman derived logical consequences from **TD13**:
"(i) An elastic material with heat conduction and viscosity is determined by the three response functions ε^*, l, q^*;
(ii) The *temperature* is given by $\theta = \theta(F, \eta) = \varepsilon^*_\eta(F, \eta)$;
(iii) The *stress* is given by $T^*(F, \eta) = \rho F \varepsilon^*_F(F, \eta)$;
(iv) The *response function l* is restricted by the dissipation inequality

$$\text{tr}\{Ll(F, \eta)[L]\} \geq 0 ;$$

(v) If the temperature relation is invertible, the response function q^* is restricted by the heat conduction inequality

$$vq^*(F, \eta, v) \leq 0 ,$$

for all vectors v, all tensors F and all scalars η and hence also

$$q^*(F, \eta, -v) = -q^*(F, \eta, v)" \quad [12.60] .$$

In the section of the paper {2.21} devoted to the case of *elastic fluid with heat conduction and viscosity*, Noll and Coleman obtained the following results:
(i) "The shear viscosity μ and the bulk viscosity $[\lambda + (2/3)\mu]$ should both be non-negative";
(ii) "The heat viscosity κ couldn't be negative" [12.61].
In August of 1963, Noll and Coleman finished the paper "Material Symmetry and Thermostatic Inequalities in Finite Elastic Deformations" {2.25}. The purpose of this work was to "investigate the general limitations placed on s (the present stress on an elastic material point) by material symmetry and thermodynamic considerations" [12.62]. Noll and Coleman showed that the results of their study within the framework of simple materials had a practical value in the classical theories of elasticity and fluid dynamics. The following axioms were formulated:

ED1: (Fundamental Constitutive Assumption of Elastic Material)

To each local configuration M there corresponds a unique value of the stress tensor $s = g(F)$, where g is the response function, taken relative to a local reference configuration M_0, F is the deformation gradient [12.63].

ED2: (Assumption of Material Objectivity)

The dependence of the stress s on the local configuration M is such that if $M_2 = QM_1$, where Q is any orthogonal tensor, then

$$s_2 = Qs_1Q^T ,$$

where s_1 and s_2 are the stresses corresponding to M_1 and M_2 respectively [12.64].

ED3: (Isotropy Group)

The group T of unimodular tensors H for which the identity

$$g(FH) = g(F)$$

holds for all F, where g is the response function relative to the reference configuration M_0, is called the *isotropy group relative to* M_0 [12.65].

ED4: (Elastic Solid)

If a material is such that for some M_0, T is a subgroup (which need not be proper) of O (the orthogonal group), then we say that the material is a *solid*. M_0 is called an undistorted local configuration [12.66].

ED5: (Elastic Fluid)

It is said that a material is a *fluid* if T is the full unimodular group U [12.67].

Noll and Coleman "investigated the limitations material symmetry imposed upon the stress of undistorted configurations of various types of aeolotropic solids" [12.68].

ED6: (Hypothesis on Crystalline Solids)

Consider a crystal which, relative to some undistorted

state M_0, has the crystallographic symmetry group (i.e. point group) G. The isotropy group T of this crystal, relative to M_0, is assumed to be the group generated by G and the inversion $-I$ [12.69].

As they managed to find out, "the 32 (known) types of (crystallographic) symmetry groups gave rise to only 11 types of isotropy groups". They investigated those isotropy groups in full [12.70]. The next step was to turn to the thermostatic inequalities. Noll and Coleman considered the first Piola-Kirchhoff stress tensor in a general form

$$t = h(F) \ .$$

ED7:   (Weakened Thermostatic Inequality or WTI)
(1) The response function h is such that, for all pairs of tensors F^*, F for which $F^* \neq F$ and $F^* F^{-1}$ is positive-definite and symmetric, one has

$$\text{tr}\{(F^* - F)[h(F^*) - h(F)]^T\} > 0 \ ;$$

(2) (WTI in an Equivalent Form)

$$\text{tr}\{(U - I)F[h(UF) - h(F)]^T\} > 0$$

for all F and all positive-definite symmetric $U \neq I$;
(3) (WTI in the Second Equivalent Form)
Since the reference configuration can always be chosen such that $F = I$, the WTI is equivalent to the requirement that

$$\text{tr}\{(U - I)[h(U) - h(I)]\} > 0$$

for all positive-definite symmetric U, no matter what reference configuration is used in defining the response function h [12.71].

Walter Noll and B.D. Coleman investigated the limitations imposed by the WTI on the properties of local configurations which gave rise to hydrostatic pressure [12.72]. Then, they demanded the response function g from the EDI (or, equivalently, h) to be continuously differentiable and studied "the restriction that the WTI imposed on the gradients of g and h" [12.73]. The scientists also found the "limitations which the WTI placed on the elastic coefficients characterizing the response of an isotropic material to an infinitesimal deformation from an undistorted state" with particular attention to the case of materials with cubic symmetry [12.74].

13   Neo-Classical Space-Time of Mechanics

W alter Noll's paper on the concept of the symmetry group of a physical system {2.40} contains a mistake with respect to the first source of the neo-classical space-time of mechanics. He made it in his report at the Delaware Seminar on the Foundations of Physics {2.30}, published in 1967. In reality, this discovery was made public for the first time in his lectures on the foundations of mechanics {2.29}, given by him at the International Mathematical Summer Center of Bressanone, Italy, in June 1965 [13.1].

Noll was one of the few who occupied themselves with the space-time problem of mechanics. From his personal experience, he came to the conclusion that "classical space-time was not well suited even for some venerable branches of mechanics", and he set himself the task to "develop another structure ... more appropriate for these branches" [13.2]. In order to achieve this, he had to restrict the use of space-time concepts, formulated in ordinary language, since "the natural language kept ... in the prison of classical space-time" [13.3]. As in numerous other cases, in the development of the new space-time concept he relied on the language of axiomatic mathematics [13.4]. The new space-time had to satisfy three main demands: (i) It had to be suitable for the mechanics of continuous media (and, given that, also for classical mechanics); (ii) It had to be compatible and to give sense to the principle of frame-indifference or objectivity [13.5]; (iii) The new structure of space-time should relate to the Euclidean space. Implicitly, Noll also added the fourth condition: he wanted to follow the approach of Hermann Minkowski to space and time, dating back to 1908 [13.6]. Noll intended to introduce mathematical concepts for neo-classical space-time in such a way that they could be interpreted physically as the results of measurements of time with clocks and of space with measuring sticks and "reflected familiar experiences with such measurements" [13.7]. He considered the concept of space-time a combination of three primitive concepts: the event-world, the distance function and the time-lapse function.

Classical Space-Time. This included: (i) The *event-world* $W = E \times E$, where E was called the *absolute (Newtonian) space*, composed of *points (locations)* x, **R** was identified with the set of real numbers and each of its elements t was called *time* of the event (x, t); (ii) the *time-lapse function* between $a, b \in \mathbf{R}$ could be defined as $\{a - b \mid a \geq b\}$ or $\{b - a \mid b \geq a\}$; (iii) the *distance function* on E was axiomatically defined as a Euclidean metric, such that E together with it built a Euclidean space [13.8].

Neo-Classical Space-Time. Keeping close to the content of the Minkowskian report of 1908, Noll proposed an *event-world* of *neo-classical space-time* to be an arbitrary set W, whose elements were called *events*. The *time-lapse function* τ on W for neo-classical space-time satisfied the following axioms:

TLF1: $\forall a, b \in W \times W \; \exists \tau(a, b) \in \mathbf{R}$, where $\tau(a, b)$ was the *time-lapse* of a and b;

TLF2: $\forall a, b \in W \colon \tau(a, b) = -\tau(b, a)$;

TLF3: $\forall a, b, c \in W \colon \tau(a, b) + \tau(b, c) = \tau(a, c)$;

TLF4: $\forall a \in W, \forall t \in \mathbf{R}, \exists b \in W \colon \tau(a, b) = t$ [13.9].

With the help of the time-lapse function τ, Noll was able to give a precise mathematical sense to the words *earlier*, *later* and *simultaneous* for pairs of events a, b from W:

TLF5: $\{a$ is *earlier* than $b\}$ iff $\{\tau(a, b) > 0\}$;

TLF6: $\{a$ is *later* than $b\}$ iff $\{\tau(a, b) < 0\}$;

TLF7: $\{a$ is *simultaneous* with $b\}$ iff $\{\tau(a, b) = 0\}$.

TLF8:   The pair (a, b) of events from W is in the *simultaneity-relation* S if $\tau(a, b) = 0$ [13.10].

The simultaneity-relation S is, as can be easily shown, an equivalence relation [13.11]. Then, it defines a partition Γ of the event-world W into non-intersecting classes of simultaneous events in terms of S:

$$W = \bigcup_{a \in \Gamma} I_a \, .$$

TLF9: $I_a = \{b \in W \mid \tau(a, b) = 0\}$ is an *instantaneous space* or *instant*.

The time-lapse function τ could then also define time-lapses between instants:

$$\tau(I_a, I_b) = \tau(c, d), \quad \forall c \in I_a, \quad \forall d \in I_b \, .$$

The *distance function* δ on W for neo-classical space-time was specified by Noll in the following axioms:

DF1: $\delta_a: I_a \times I_a \to \mathbf{R}$, where $a \in \Gamma$;

DF2: $\forall a \in \Gamma: \delta_a$ is an Euclidean metric on I_a, which endows I_a with the structure of an Euclidean point space;

DF3: $\forall a \in \Gamma: dim \, V_a = 3$ [13.12].

The discovery of neo-classical space-time led Noll to an important philosophical conclusion on the role of the principle of objectivity or frame-indifference [13.13]. The core of these principles, as he wrote, is that "only the neo-classical event-world W itself has physical significance, and not any representation $W = E \times \mathbf{R}$ of W in terms of an absolute space E. In other words, there is no such thing as 'physical space'" [13.14].

Noll's neo-classical space-time received a high evaluation in the monograph "Foundations of Physics" (1967) by Mario Bunge, who wrote there: "The first exact relational theory of time was proposed only recently (Noll, 1967) – which goes to show the sad state of neglect of foundations research. (For previous unsuccessful attempts see Carathéodory, 1924, and Reichenbach, 1924). It is a theory of *universal time* in the sense that it assumes a single time for every place and every reference frame; it will be abbreviated UT. This theory specifies a relational, not an absolute, time concept because it analyzes time as a relation among pairs of events on the one hand, and numbers on the other. Now, since time lapses not only ride on events, but are relative to reference frames, UT may hold for every single frame but it fails to state the timing relations among events referred to different frames. We shall therefore relativize it further, namely to the frame concept, building along its lines a theory of *local time* (LT)" (p. 95).

14 Inhomogeneities in Simple Bodies

In 1963, Walter Noll finished a preliminary version of his treatment of the continuum theory of dislocations. The tasks of this study were: (i) To generalize and to give a rigorous foundation of the subject; (ii) To make this theory free from *ad hoc* assumptions on the geometrical structure of the body. The power of Noll's approach lay in the fact that "once a constitutive equation such as to define a materially uniform simple body was laid down, the geometric structure in the body was determined". This structure wasn't an assumption of the theory of dislocations, but could be inferred from it [14.1]. The preliminary manuscript of 1963 didn't satisfy Noll. It remained unpublished until 1965, when he included the summary to §34 of his joint fundamental treatise on non-linear field theories of mechanics {1.2}. However, he distributed several copies of the manuscript among his close friends, including C.A. Truesdell, C.-C. Wang, R. Toupin, and probably some others [14.2].

In 1967, Noll published the final version of his theory of dislocations and made a survey report on it at the IUTAM Symposium on the Mechanics of Generalized Continua, held in Freudenstadt, West Germany [14.3]. In the following exposition of his theory of inhomogeneities in simple bodies it would be profitable to discuss also the boundaries of his *method of conceptualization*.

In the version of 1965, Noll used the coordinate-free approach: he exchanged the usual reasoning in terms of coordinates of conceptualized objects for that in terms of classes of objects. Each class could be defined mathematically in the standard way, once an appropriate relation of equivalence was found. Moreover, such classes could be characterized through one of its members which was to be called the *reference element*. Noll began with the postulate that he dealt with the simple materials, already known to us from Chapter 9. A *local reference configuration* was an example of equivalence classes. From the definition of simple material, he passed to a subclass of it, which he called a *uniform reference*. A fixed uniform reference gave a simple body two geometric structures: a *Riemannian metric* and a *linear connection*, which were both purely mathematical concepts. Noll showed that they both defined the *homogeneity of a simple body*: (i) A simply connected materially uniform body was said to be *homogeneous* iff it had a symmetric material connection; (ii) Under the assumptions that the body was homogeneous and simply connected, its fixed local reference was undistorted, and the curvature of the Riemannian metric vanished, the body had to be homogeneous [14.4]. He demonstrated that his abstract theory had a common physical content, at least in the case of an *aeolotropic body*, where his concept of curvilinear aeolotropy generalized in the natural way the theory of hyperelasticity by A.E. Green and J.E. Adkins [14.5]. The term "inhomogeneity" didn't appear in {1.2}. Having made the concept of homogeneity strictly mathematical, Noll moved on in his theory to search for more and more sophisticated necessary and sufficient conditions for it. However, he soon stopped, the main reason being that his treatment had abandoned the sphere of the language of mechanics and migrated entirely to that of pure mathematics, most of which could hardly, if at all, be translated back. Thus, we come to the first boundary of Noll's conceptualization: *a mathematical theory of physical phenomena is restricted due to the properties of the mapping between the conceptual language of physics, based on intuition and experiment, and the language of abstract mathematics.* "Good theorems", according to C.A. Truesdell, were such that, for the given conceptual scheme, they could cover the whole field within this boundary due to the given task of research [14.6].

Let's turn to the principal publication of Noll on the theory of inhomogeneities {2.31} and its summary in {2.32}. From the list of references to the first publication, one can deduce that Noll's conceptualization has a second boundary: *it is limited by the contemporary state of art in the field* [14.7]. The content of the theory of dislocations was taken by Noll from the publications of E. Kröner of 1960 and A. Seeger of 1961 [14.8]. Excluding a paper of C.-C. Wang, Noll's other references were six publications of his own and a book of S. Lang on differentiable manifolds, which served as a source of "modern differential geometry" [14.9]. The above-mentioned second boundary of Noll's conceptual approach didn't allow him to base his theory of dislocations on neo-classical space-time (see Chapter 13), and he had to satisfy his ambition with a remark that "the considerations of this paper could be adapted" to new space-time [14.10]. The preliminary mathematical tools of the theory of inhomogeneities included standard definitions from the theory of linear transformations, group theory and differential geometry "as they came to be needed, tailored to the requirements of the intended applications" [14.11].

The physical concepts introduced in the first three parts of this paper were that of a local deformation, a material point, a body's configuration, a continuous body, and, finally, of a local configuration at a material point, which he presented as the key concept for his theory [14.12]. In {2.31}, Noll made *a remarkable generalization of the concept of simple body* (see Chapter 9), which covered not only its mechanical local characteristics, but also that of heat capacity, electrical conductivity, magnetic response, optical refraction, chemical composition, color, and all other thinkable local physical characteristics. He wrote in {2.32}: "The theory of simple bodies deals only with such local characteristics. A given set of physical phenomena is then governed by appropriate physical responses at the various material points of the body. The term 'simple' expresses the assumption that deformations, whose gradient is the identity at a given material point, do not alter the physical response at that point with respect to the set of phenomena under consideration. Roughly speaking, a body is simple if only first spatial gradients occur in the constitutive description of its physical properties. A simple body is called materially uniform if the physical response is the same at all points of the body" (p. 239).

Mathematically, a *simple body* was re-defined by Walter Noll through the primitive concepts of a response descriptor and a continuous body: "Let R be a set, whose elements we call response descriptors. A continuous body B of class $\mathbf{C}^p$ will be called a simple body with respect to R if it is endowed with a structure by a function G which assigns to each material point $X \in B$ a mapping

$G_X : C_X \to R$. The value $G_X(C_X)$ is the response descriptor of the material at X in any configuration γ of B such that $\nabla\gamma(X) = C_X$" [14.13]. Noll remarked that the mappings G_X were subject to "restrictions imposed by general physical principles such as the principle of frame-indifference and the principle of dissipation" [14.14].

The next step of Noll's conceptualization was the introduction of a material isomorphism, which enabled the comparison of simple materials at different points of the body in terms of the function G_X [14.15]. The concept of material isomorphism allowed Noll to narrow the class of all simple bodies to its subclass of materially uniform simple bodies, defined by the following property: "A simple body B was said to be materially uniform if the material at any two of its points was the same" [14.16]. The term "material uniformity" was assigned by him to a function which put into correspondence a pair of material points of the body with a material isomorphism between them [14.17].

Noll called a *reference* for body B such a function of it, whose values were local configurations. He specified a connection between the concepts of reference and material uniformity: every uniform reference K, for which the function $\Phi(X, Y) = K(X)^{-1}K(Y)$ was a material isomorphism, determined a material uniformity. He formulated and proved a necessary and sufficient condition for a reference to be uniform in terms of the function G_X, mentioned above in the definition of a simple body: a reference was *uniform* iff there existed a response function H_X of the body relative to the uniform reference K, of the form $H_X = G_X(FK(X))$, $X \in B$; F was a member of the group of invertible linear transformations [14.18]. Noll then defined "the *isotropy group* of the body B relative to the uniform reference K" as a subgroup of linear invertible transformations of a three-dimensional vector space, satisfying the property [14.19]:

$$g_K = K(X)g_X K(X)^{-1} .$$

He proved a theorem which established relations "between two uniform references and the corresponding response functions and isotropy groups" [14.20]. He extended his theory of isotropy groups and obtained in the natural way the concepts of undistorted uniform reference, uniform isotropic body and uniform solid body.

He devoted the next three sections of {2.31} to the "mathematical prerequisites necessary to describe the local behavior of material uniformities and uniform references that possessed a degree of smoothness" (p. 2). These mathematical concepts were ordered by Noll into groups according to the sections of the paper: (i) *vector and tensor fields*: he introduced here mappings of class $\mathbf{C}^r$, functions (scalar fields) on B, vector and tensor fields on B, tangent vector fields, intrinsic tensor fields, and F-linear mappings; (ii) *relative gradients and brackets*: this included concepts of gradient relative to a configuration, gradient relative to a reference, reference of class $\mathbf{C}^r$, bracket of two tangent vector fields; (iii) *affine connections, torsion, curvature*: affine connection of class $\mathbf{C}^{r-1}$, frame of class $\mathbf{C}^r$, components of connections, Cartan-torsion of connections, Riemann-curvature of connections, first Bianchi identity [14.21]. These mathematical concepts show that Noll's conceptualization possessed a third boundary: *it didn't mean an introduction of new mathematics, but only the use of its modern forms, adapted to particular cases.*

The next principal concept of Noll's theory of inhomogeneities was *material connection*, which he introduced in terms of affine connection and material uniformity. This definition allowed him to prove a theorem that material connections should have zero Riemann-curvature. In order to define the concept of inhomogeneity, Noll had to consider first the intuitively clear concept of *homogeneity for simple bodies*: "A simple body B is called locally homogeneous if every $X \in B$ has a neighborhood N that is homogeneous" [14.22]. In {2.32}, one can find Noll's comments on the concepts of the homogeneous and inhomogeneous body in the language of physics: "We say that the body is homogeneous if it admits homogeneous uniform references. ... Intuitively, if a body is homogeneous, we can view it in a configuration such that all parts of the body respond in the same way. If the body is inhomogeneous but materially uniform, we must first cut it into infinitesimal pieces that do not fit together before we can make all parts respond in the same way" (p. 243). Mathematically, the concept of *inhomogeneity* (for material uniformity) was defined as the Cartan torsion of the material connection associated with a material uniformity of class $\mathbf{C}^{p-1}$. Noll added to this a natural definition of the *inhomogeneity relative to a uniform reference*. He then proved a theorem which specified the interconnections between the concepts of homogeneity and inhomogeneity.

The next step in Noll's theory was an introduction of two "technical" concepts, called a Riemannian connection relative to uniform reference, and a contortion of a uniform reference. The rest of the paper {2.31} on the theory of dislocations was devoted to the applications of the introduced conceptual tools. He analysed first the case of *contorted aeolotropy*, which was, in fact, a generalization of the known curvilinear aeolotropy from the "classical" theory of dislocations. He wrote: "A uniform reference K of class $\mathbf{C}^{p-1}$ is called a state of contorted aeolotropy if there exists a configuration γ such that the tensor field $Q = (\nabla \gamma) K^{-1} \in \mathbf{L}_B^{p-1}$ has orthogonal values" [14.23]. In {2.32}, Noll described the *physical sense of contorted aeolotropy*, introduced a method of manufacturing a body in this state, and specified its connections with the usual curvilinear aeolotropy. He wrote: "Take very many thin sheets of a homogeneous material and bend them into cylindrical shape in such a way that they can be stacked snugly. Then glue them together with a homogeneous glue. One thus obtains a body in a certain configuration γ. If we consider small pieces of the body at material points X and Y, we see that these pieces can be brought into alignment by rotating one of them. For the body we have just described, we can introduce a cylindrical coordinate system and choose the orthogonal tensor field Q in such a way that $Q(X)e_i(X)$ is independent of X when $e_i(X)$, $i = 1, 2, 3$, are unit vectors pointing in the direction of the coordinate lines. When an orthogonal curvilinear coordinate system with the property just mentioned exists, we say that the resulting uniform reference is a reference of curvilinear aeolotropy. Not all references of contorted aeolotropy are also of curvilinear aeolotropy. An example is given by taking many very thin fibers of homogeneous material, twisting them together as in a rope, and then glueing them together" (p. 244). The concept of contorted aeolotropy could also be defined, according to Noll, in terms of the curvature of the Riemannian connection relative to the uniform reference K [14.24].

Noll proved three results related to materially uniform bodies:

(i) "If the isotropy groups of a materially uniform simple body B were discrete, then B had at most one continuous material uniformity Φ" [14.25];
(ii) "If a uniform isotropic body had an undistorted state of contorted aeolotropy, it was homogeneous" [14.26];
(iii) For a uniform simple body with a distinguished class of uniform references with a special property, the Riemannian connection and its curvature were characteristics of the body [14.27].

Noll obtained an appropriate version of Cauchy's equation of balance in its general and some practically used forms, which "expressed the fact that the forces acting on every part of a given body B should add to zero" [14.28]. This last section was of much practical importance, and it allowed an applied mathematician or an engineer to use the abstract theory of inhomogeneities in simple bodies for his everyday needs.

15 Fit Regions and Contact Interactions

In 1985, during his stay in Pisa, Italy, Walter Noll made his first contact with Epifanio G. Virga, who later became his most important collaborator after B.D. Coleman and H.D. Dombrowski [15.1]. In October-November 1985, Virga was staying at the Department of Mathematics of the Carnegie-Mellon University on the invitation of Professor W.O. Williams. He was working at that time on a paper devoted to the "unstable equilibrium configurations of fluids which were not removed by strengthening surface tension" [15.2]. Williams offered Virga an opportunity to speak on his research at a department colloquium. Virga remembered: "Walter (Noll) was in the audience. After the talk he said to me that he had attempted to find the simplest class of sets which would form a 'material universe' according to his definition of 1959, but he was stuck in some difficulties with sets of finite perimeter. He ended by proposing to work together on the problem. I was happy to accept" [15.3]. Noll and Virga collaborated first in Pittsburgh until the end of Virga's stay. They continued by correspondence. However, the problem was too difficult for them both. Noll could not read in Italian the works of Ennio De Giorgi on sets with finite perimeter, which were of immense importance for their work. On the other hand, Virga was only a "beginner" at that time, as he wrote himself, and, obviously, he was insufficiently qualified to help him. In June 1987, Noll went again to Pisa. He and Virga managed to obtain the necessary consultations from De Giorgi himself, which resolved their difficulties. At the beginning of August 1987, Noll and Virga finished a paper entitled "Fit Regions and Functions of Bounded Variation" {2.47}, which appeared in 1988 [15.4].

Noll's attention to fit regions in the middle of the 1980s was triggered by a remarkable success in this direction, achieved by M.E.

Gurtin, W.O. Williams and W.P. Ziemer and published in 1986 [15.5]. By that time, Noll possessed only a set of requirements which any fit region should satisfy: "First, the set of all fit subregions of a given fit region should satisfy the axioms of a 'mathematical universe'. ... Second, the class of fit regions should be invariant under transplacements, which should include adjustments to fit regions of smooth diffeomorphisms from one Euclidean space to another. Third, each fit region should have a surface-free boundary for which a form of the Integral-Gradient Theorem (also called Gauss-Green theorem) should be valid" [15.6]. Noll included these conditions in a precise mathematical form in his paper {2.45}, published in 1986. At the same time, he considered the above-mentioned results of Gurtin, Williams and Ziemer as not at all final. Though they met three requirements of fit regions, the obtained class seemed to be "unnecessarily large" [15.7]. Moreover, Noll rejected the thought that Gurtin, Williams and Ziemer were first in relation to the idea of usage of sets of finite perimeter for a definition of fit regions, and dismissed their contribution as not detailed enough [15.8]. C.A. Truesdell stood by Noll, and the second edition of the first volume of Truesdell's textbook "A First Course in Rational Continuum Mechanics", contained the theory of fit regions elaborated by Walter Noll and E.G. Virga, not by Gurtin, Williams and Ziemer [15.9]. Noll proclaimed the problem of specifying "fit regions" to be one of the two issues necessary to "arrive at a fully satisfactory mathematical foundation of continuum physics" [15.10]. His vague idea on how to find a class of fit regions under the mentioned conditions was to replace "the concepts of 'closure' and 'interior' by the concepts of 'essential closure' and 'essential interior' from the geometric integration theory" [15.11]. As necessary conditions for a theory of fit regions, Noll considered its simplicity and practical value for an engineer [15.12]. In {2.47}, Noll and Virga could give only a slight difference of their class of fit regions from that of Gurtin, Williams and Ziemer. The two scientists wrote: "It is the purpose of this paper to define a class $\underline{Fr}$ of fit regions ... which satisfies the three requirements mentioned and *cannot easily be made any smaller*. We believe that the class $\underline{Fr}$ should be used as a basis for mathematical theories of continuum physics" (pp. 2–3). As far as notation and terminology are con-

cerned, the paper {2.47} is a masterpiece, written in the spirit and letter of Noll's fundamental treatise on finite-dimensional spaces. The
first simple exposition of the Noll-Virga theory of fit regions was
made by Truesdell in 1991 [15.13].

Noll and Virga began their article {2.47} with a counter-example,
showing that "sets with piecewise smooth boundaries" were generally unsuitable for the role of fit regions, i.e. "sets to be occupied by
continuous bodies and their subbodies". Thus, Noll acknowledged a
deficiency in his concepts of a continuous body, discussed in Chap-
ters 9 and 14. If $A = \{(a, b) \mid a, b \in \mathbf{R} \wedge a \in (0, 1) \wedge b \in (0, 1)\}$
and $B = \{(a, b) \mid a, b \in \mathbf{R} \wedge a \in (0, 1) \wedge b \in (-1; \exp(-1/t^2)$
$\sin(2/t))\}$ then ∂A and ∂B were piecewise and of the class $\mathbf{C}^\infty$.
However, $A \cap B$ "consisted of *infinitely many* pieces and could cer-
tainly not be regarded as a set with a piecewise smooth boundary"
[15.14].

Noll and Virga gave the following definition of a *fit region*: "We
say that a subset D of E (a three-dimensional Euclidean point-space)
is a fit region in E if it (i) is bounded; (ii) is regularly open; (iii) has a
finite perimeter; (iv) has a regular boundary. The set of all fit regions
in E will be denoted by $\underline{Fr}(E)$ and the class of all fit regions (in all
Euclidean spaces) by $\underline{Fr}$" [15.15]. What is the role of sets of finite
perimeter in the theory of fit regions? The six general requirements
of candidates for the role of a fit region were formulated by Noll in
1986 [15.16]. The condition (i) could be violated (and thus it wasn't
necessary) in the continuum physics, where it is sometimes useful to
consider also unbounded bodies, for example, as large as half of the
Euclidean plane. Noll and Virga indicated a recipe to expand their
theory of fit regions for these cases: "We believe that such bodies
should be considered as 'improper bodies' and should be treated
in terms of proper bodies in a way which parallels the treatment
of improper integrals in terms of proper integrals in elementary
analysis" [15.17]. The condition (ii) was taken from {2.45}, where
Noll also pointed out a natural substitute for it:

N1:   "N (a fit region) is a subclass of $\underline{Regel}$ (the class of
            all closed sets that coincide with the closure of their
            interior). It makes no substantive difference whether one

uses the class <u>Regel</u> or <u>Regop</u> (the class of all open sets that coincide <u>with the interior of their closure</u>), because the process of taking the closure is a natural one-to-one correspondence from <u>Regop</u> onto <u>Regel</u>, its inverse being the process of taking <u>the interior. In</u> some contexts it is easier to work with the class <u>Regop</u>" (pp. 9–10, 12).

The condition (iv) could be traced back to the following property of the class N in ({2.45}, p. 12):

N4:   "It is possible to define an area-measure area on $\mathrm{Bdy}\, C$ (C is an element of N, and $\underline{\mathrm{Bdy}}$ denotes the operation of taking the boundary) in such a way that for each $D \in N$ with $D \subset E$ and

$$\underline{\mathrm{Clo}}\,\underline{\mathrm{Int}}(C \cap D) = \emptyset\,,$$

the subset $C \cap D$ of $\underline{\mathrm{Bdy}}\, C$ is area-measurable" [15.18].

Noll preferred the condition (iv), since "given a subset of a Euclidean space, it was usually easier to decide whether the boundary was negligible than whether the boundary had finite area-measure" [15.19]. The condition (iii) was the decisive one in the Noll-Virga theory of fit regions, and, in different combinations with other conditions, it allowed them to obtain the rest of Noll's requirements for a fit region as theorems. In {2.45}, Noll demanded for the elements of the class N:

N3:   "If $C, D \in N$ and C and D are both subsets of the same Euclidean space, then $\underline{\mathrm{Clo}}\,\underline{\mathrm{Int}}(C \cap D) \in N$, $C \cup D \in N$, and $\underline{\mathrm{Clo}}(C \backslash D) \in N$" (p. 12).

In {2.47}, he and Virga obtained it as a theorem: "If A and B are fit regions in E, so are the intersection $A \wedge B$, the joint $A \vee B$ (= $\mathrm{IntClo}(A \cup B) \ldots$, and the difference-region $A \triangle B$ (= $\underline{\mathrm{Int}}(A \backslash B) \ldots$" (p. 15). In {2.45}, Noll also imposed a demand on the class N:

N2:   "N is invariant under displacements" (p. 12).

In {2.47}, Noll and Virga proved it as the following theorem: "Let $f : E \to E'$ be a $\mathbf{C}^1$-diffeomorphism from the given Euclidean space E onto an Euclidean space E'. Then the image under f of every fit region in E is a fit region in E'" (p. 16). The last two requirements of Noll's in {2.45} for the candidates to be a fit region were:

N5:   "It is possible to define an *exterior unit normal* function

$$\underline{n}_C \colon \mathrm{Bdy}\, C \to V$$

(V is the Euclidean vector space of E), which assigns to each point of $\mathrm{Bdy}\, C$ the unit vector normal to the 'surface' $\mathrm{Bdy}\, C$ and directed away from C. The values of $\underline{n}_C$ are unique except perhaps on a subset of $\mathrm{Bdy}\, C$ that has area-measure zero. If f is a continuous function on an area-measurable subset S of $\mathrm{Bdy}\, C$, it is possible to form the area-integral

$$\int_S f\underline{n}_C \mathrm{d}(\underline{\mathrm{area}}) \in V \ ;$$

N6:   If the set C is bounded then the Integral-Gradient Theorem (often called Green-Gauss theorem) applies. We mean by this that for every differentiable function f on C whose gradient ∇f is volume-integrable, we have

$$\int_C \nabla f \mathrm{d}(\underline{\mathrm{vol}}) = \int_{\mathrm{Bdy}\, C} f\underline{n}_C \mathrm{d}(\underline{\mathrm{area}})" \ \text{(pp. 12–13)}.$$

With the help of the concepts "outer unit normal", "reduced boundary" and "outer-normal field", Noll and Virga obtained **N5** and **N6** as an integral-gradient theorem in {2.47}: "For every continuous function $f : \underline{\mathrm{Clo}}\, D \to \mathbf{R}$ (D is a fit region) for which $f_{|D}$ is differentiable and $\nabla f_{|D}$ integrable on D, we have

$$\int_D \nabla f_{|D} \mathrm{d}v = \int_{\underline{\mathrm{Rby}}\, D} f\underline{n}_D \mathrm{d}a \ ,$$

$\underline{\mathrm{Rby}}\, D = \{x \in E \mid$ there exists an outer normal to D at $x\}$" (p. 17).

They finished the paper with a section devoted to some useful "counter-examples" of sets, which satisfied only part of the conditions adherent to the fit regions. They also included an example of fit regions according to the theory of Gurtin, Williams and Ziemer [15.20]. Together with Noll's paper {2.45}, the Noll-Virga theory of fit regions in {2.47} had to serve as a basis of a theory of *contact interactions*, which was the second and last issue "to arrive at a fully satisfactory mathematical foundation of continuum physics" [15.21].

In 1987, Noll met in Pisa with Maurizio Vianello, a mathematician from Milan, who was also very interested in the construction of a non-trivial theory of edge interactions. However, Vianello wanted to make this construction along the lines of R.A. Toupin's theory of second-grade materials, dating back to the 1960's [15.22]. Obviously, it was an old-fashioned and backward theory for 1987. During the meeting between Noll and Vianello, at which Virga was also present, they decided to collaborate on the issue and worked out several basic ideas [15.23]. Curiously enough, the first success here fell on the side of Virga, who managed in the summer of 1987 to obtain "a formula for simple edge interactions" and to sketch "an argument to prove it" [15.24]. However, Vianello "was rather sceptical about the approach" used by Virga. The reason for this was Vianello's inclination to Toupin's early approach in the continuum mechanics of materials. They couldn't come to an agreement, and Virga, a little disappointed, flew to Baltimore, where he received a temporary position at the Johns Hopkins University [15.25]. In 1987-1988, Virga went twice to Pittsburgh for a week to discuss his ideas on edge interactions with Walter Noll, who was working at that time on a general theory of contact and separation of continuous bodies. During these two weeks, Virga was lodging at Noll's house in Pittsburgh, occupying the room of his son Peter. In June 1989, Noll returned to Pisa, and he and Virga completed the paper "On Edge Interactions and Surface Tension" {2.48}. They used most of the time "to remove subtle contradictions from the assumptions of their theory" [15.26]. The paper reached the Editorial Office of the Archive for Rational Mechanics and Analysis on August 1, 1989, and was published in 1990 [15.27].

One of the tasks of the paper was for the first time "to bring into the open the systems of forces that gave rise to distributions of forces over edges" [15.28]. Noll and Virga introduced the main mathematical tools they intended to use in the theory of edge interactions. These were the mathematical concepts of a *region* (in a three-dimensional Euclidean space E), *surface, border, side, edge, vertex, vertex-point, adjacent edge* etc. In order to make their theory non-trivial, they decided to consider only "regular" concepts: for example, *regular regions and surfaces*. The bounded connected $\mathbf{C}^2$-manifold A was called "regular" if its closure had a finite partition P with special properties: (i) Every element of P was a connected $\mathbf{C}^2$-manifold; (ii) A was itself an element of P; (iii) For every element $B \in P$ there existed a continuous extension of the mapping $f_{|B}$ to the closure of B, where $f_{|B} : B \to \underline{\operatorname{Lin}} V$, V being the translation space of E, and $f_{|B}$ was a symmetric idempotent mapping for every $b \in B$ such that

$$f_{|B}(b)^2 = f_{|B}(b); \quad [f_{|B}(b)]^T = f_{|B}(b) .$$

The range of $f_{|B}(b)$ coincided with the tangent space to B at b for all $b \in B$; (iv) For every $B \in P$ and every $C \in P$ with $C \subset$ [Closure $(B)\backslash B$] and $\underline{\dim} B = \underline{\dim} C + 2$, there were exactly two pieces $D \in P$ such that $D \subset$ [Closure $(B)\backslash B$], $C \subset$ [Closure (D) $\backslash D$]. A finite partition of the closure of A with the properties (i)–(iv) was called a "regular partition" for A [15.29]. Regular manifolds of the theory had to possess a special border – "reduced border", a "union of all regular manifolds of dimension ($\underline{\dim} A - 1$) that were included in [Closure $(A)\backslash A$]" [15.30]. Every "regular region" was then also a fit region and its "reduced border" was a subset of the "reduced boundary" from {2.47}. Noll and Virga specified the concept of the outer unit normal to a regular manifold at its point, and they explained its properties in a proposition. In the traditional Walter Noll style, they gave a definition of a *differentiable mapping* of class $\mathbf{C}^1$ from the regular manifold to a finite-dimensional linear space and of a *surface-divergence* [15.31]. Noll and Virga gave one more special definition, which allowed them to formulate a corollary of the classical Stokes theorem: "Let B be a surface of class $\mathbf{C}^2$. We say that a function $h : S \to$

V is *tangential* if $h(x) \in \underline{\text{range of}} f_{|B}(b)$. We say that a function $H : B \to Lin(V, W) = \{$the set of all linear mappings from V to $W\}$ is tangential if $\{[\underline{\text{range of}} f_{|B}(b)]^{\perp} \subset \underline{\text{Null}} H(x)\}$ for all $x \in B$" [15.32]. The corollary of the Stokes theorem was called "Surface-Divergence Theorem": "Let B be a regular surface. For every function $h : \underline{\text{Closure}} B \to V$ that is of class $\mathbf{C}^1$ and tangential we have

$$\int\limits_{B} \underline{\text{div}_B} h \mathrm{d}a = \int\limits_{\substack{\text{regular} \\ \text{border of } B}} (h v_B) \mathrm{d}l \, ,$$

where v_B is the outer unit normal function of B" [15.33]. The two mathematicians formulated a corollary of the Surface-Divergence Theorem for tangential functions [15.34].

It was a general defect of Noll's axiomatization of continuum mechanics which was repeated in {2.48}: the violation of the axiom of impenetrability of the medium. Noll and Virga defined a "contact" of two bodies as a set of border points which belonged to each of them! Due to the concept of "reduced boundary", Noll and Virga were able to introduce the concept of "reduced contact", and the contacting parts of the bodies were then said to be "in smooth contact" [15.35]. They brought into play the concept of *outer unit normal*: "If B and B' were regular surfaces, one should say that B and B' were *in cusped contact* at a given x, belonging to the intersection of reduced boundaries of B and B' if the outer unit normals to them at x coincided. Obviously, B and B' could be *in everywhere-cusped contact* along a general curve of them or *in nowhere-cusped contact* there" [15.36]. According to the definition of cusped contacts, Noll and Virga reduced the generality of their theory to "biregular regions", i.e. regular regions having "biregular partition". The last concept received the following clarification: a biregular partition was a regular partition such that for each of its edges the two sides adjacent to it were either in everywhere-cusped contact (cusped edge) or in nowhere-cusped contact along this edge [15.37].

It isn't difficult now to reconstruct the further reasonings of Noll and Virga, who proceeded with definitions of concepts of *biregular contact, simple surface contact, simple edge contact, and appropriate partition and section of biregular region* [15.38]. The next step in the conceptualization of the theory of edge interactions was to define a *join, an "inner" exterior, parts*, and *interior parts* for a biregular bounded region and its sections [15.39]. Noll and Virga used the axiomatical theory of forces. They adapted it to the case under study and introduced the mathematical concepts of *contact-interaction, quasi-balanced* and *skew contact interactions* [15.40]. The paper {2.48} is ingenious since all further reasoning therein was done analogically to a mathematically trivial theory of wedges or open sectors in an open disk in the Euclidean plane. Noll and Virga defined a material universe of wedges on the disk and found a solution to the problem of specifying all interactions of wedges which had common sides (otherwise they were to be zero) [15.41]. The transformation of the theory of contacts among the wedges to the case of interactions among the parts of a biregular region began with a classification of all six typical cases of contacts, which were characterized both in terms of symbols and graphically [15.42].

Noll and Virga based the further theory on four assumptions, which, as they remarked, were lacking "more transparent and natural physical interpretations" [15.43]. The *first assumption* stated, roughly, that the edge interaction along a curve of contact of two parts could be obtained by integrating wedge interactions on the cross-sections along this curve [15.44]. The *second assumption* said: for every biregular part of the region under study there existed a continuous function from that part of regular border which didn't belong to the boundary of the region, to a finite-dimensional linear space, which was the integral density of a contact interaction between this biregular part and another disjoint section of the region, whose properties were specified as three cases [15.45]. Three technical results were derived from these assumptions as a consequence of the coupling of them with the conceptualized contact interactions.

After an introduction of the *orientation* for regular surfaces, Noll and Virga expanded these results to the field of *oriented regular surfaces*. The power of their concept of orientation lay in the fact that it allowed them to differentiate between the *pure surface* and *pure edge interactions* [15.46]. The *third assumption* and its version with a clear mathematical content were introduced for tackling the resultant actions. With their help, Noll and Virga obtained a technical result, equivalent to the corollary of the surface-divergence theorem [15.47]. The technically essential fourth assumption was formulated in connection with the quasi-balanced interactions [15.48]. Two theorems which followed contained formulas for the quasi-balanced contact interactions for which the four assumptions were satisfied [15.49]. The last assumption of the theory of edge interactions strengthened the condition of quasi-balance and gave an elegance to it, differentiating between the surface action of the environment on the biregular region and the actions of its adjacent parts [15.50]. As corollaries to it, Noll and Virga derived another two nontrivial, good theorems of technical character and calculated several examples, using some results from Virga's works on the mechanics of surface tension [15.51].

16 Foundations of Special Relativity

In 1964, Walter Noll published an axiomatic of the Minkowskian chronometry, "the structure of space-time appropriate to Einstein's special relativity" [16.1]. The original work of Hermann Minkowski, "Raum und Zeit" (1908), included a programmatic statement that "the assumption of the group G_c for the physical laws leads nowhere to a contradiction, it is essential to carry out a revision of the whole of physics founded on the assumption of this group" [16.2]. It can be concluded today that Minkowskian chronometry has, in fact, influenced mechanics, but it hasn't led to a revision of it [16.3]. Noll's study of Minkowskian chronometry stimulated his search for a new appropriate space-time structure of classical mechanics, which ended with the discovery of its neo-classical space-time [16.4]. The report of Minkowski was valuable and dear to Noll due to the group-invariant approach to the governing equations of Newtonian mechanics and due to Minkowski's strict differentiation of the coordinate system from the frame of reference [16.5].

Noll's investigation of the axiomatical structure of Minkowskian chronometry was evidently triggered by two reports which he heard and discussed at the International Symposium on the Axiomatic Method with Special References to Geometry and Physics, held at the University of California on December 26, 1957–January 4, 1958. The first report, entitled "Axioms for Cosmology", was done by a British mathematician, Arthur Geoffrey Walker, who proposed foundations for relativistic cosmology in terms of an undefined basis, which involved the concepts of "particles", "light signals" and "temporal order" [16.6]. According to a respected authority on the subject, John W. Schutz, "Walker's axiom system was not developed sufficiently to describe Minkowski space-time and was, in fact, restricted to sets of relatively stationary particles, but it succeeded in clarifying many kinematic concepts, especially those of 'particle', 'light signal', 'optical line' and 'collinearity'" [16.7]. Walker's axiomatic dated back to at least 1948, while another axiomatical system for relativistic kinematics, presented in the report of an American mathematician, Patrick Suppes, was an entirely fresh one [16.8]. Noll's interest in Minkowskian chronometry and his mathematical talent to grasp new ideas and to improve them were testified to by Suppes, who wrote in this report:

The single axiom we require is embodied in the following definition.
Definition 3. A system $\aleph = \langle X, \mathfrak{I}, c \rangle$ is a *collection of relativistic frames* if and only if for every x, y in X (the set of physical space-time points), whenever x and y lie on an inertial path with respect to some frame in $\mathfrak{I}$ (an inertial space-time frame of reference), then for all f, f' in $\mathfrak{I}(:)$

$$I_f(x, y) = I_{f'}(x, y) \ .$$

I originally formulated this invariance axiom so as to require that equation ... hold for *all* space-time points x and y, that is, without restricting them to lie on an inertial path (with respect to some frame in $\mathfrak{I}$), Walter Noll pointed out to me that with this stronger axiom no physically motivated arguments of the kind given below are required to prove that any two frames in $\mathfrak{I}$ are related by a linear transformation; a relatively simple algebraic argument may be given to show this [16.9].

The elegance and beauty of Suppes' axiomatic lay in the fact that he succeeded in showing that "the single assumption needed for relativistic kinematics was that all observers at rest in inertial frames obtained identical measurements of relativistic distances along inertial paths when their measuring instruments had identical calibrations" [16.10].

For his axiomatical system of Minkowskian chronometry, Noll could have found support not only in the works of Hermann Minkowski, Arthur Geoffrey Walker and Patrick Suppes. In 1911–1914, a British mathematician and physicist, Alfred Arthur Robb, elaborated and published an axiomatic of a new concept of space-time [16.11]. Robb wrote: "We have shown how from twenty-one postulates involving the ideas of *after* and *before* it is possible to set up a system of geometry in which any element may be represented by four coordinates x, y, z, t. Three of these, x, y, z, correspond to what we ordinarily call space coordinates, while the fourth corresponds to time as generally understood. Since, however, an element in this geometry corresponds to an instant, and bears the relations of *after* and *before* to certain other instants, it appears that the theory of space is really a part of the theory of time. Spatial relations are to be regarded as the manifestation of the fact that the elements of time form a system in conical order: a conception which may be analysed in terms of the relations of *after* and *before*" [16.12]. Robb's contribution is remarkable since he proved 206 theorems and also showed the consistency of his axiomatical system. However, he couldn't avoid a certain amount of redundancy, which he "permitted to remain" [16.13].

It is typical of Noll's mathematical works that he has always attempted to see the topic (here Minkowskian chronometry) with his own eyes and avoided to blind himself with other axiomatics which had already been constructed. This can be directly concluded from his list of references to the paper {2.24}, which included the above mentioned report of Suppes, a volume of N. Bourbaki on algebra, a book on relativity theory and P. Halmos' textbook on finite-dimensional spaces. Noll's axiomatic for Minkowskian chronometry was slightly connected to that of Suppes: Noll considered it a possible answer to a problem posed in the work of Suppes [16.14].

Noll supported the alleged priority of Albert Einstein's chronometry before Hermann Minkowski's "Raum und Zeit" [16.15].

One of the general advantages of Noll's axiomatic is the lack of restrictions on the dimensionality of the spaces, which can even be infinite-dimensional. Another advantage is the use of "direct coordinate-free methods" [16.16].

Noll began his axiomatic of special relativity with the introduction of a real vector inner product space V and its subspaces:

(i) $V_+ = \{v \mid v^2 > 0 \text{ or } v = 0\}$, called "space-cone";

(ii) $V_- = \{v \mid v^2 < 0 \text{ or } v = 0\}$, called "time-cone";

(iii) $V_0 = \{v \mid v^2 = 0\}$, called "signal cone";

the index of V was the maximal dimension of the time-like subspaces of it.

This allowed him to prove a generalization of the famous inertia theorem of J.J. Sylvester in the form:

Theorem 16.1. If U is a time-like subspace of maximal dimension

$i = ind\ V$ and if $i < \infty$,

then the complement $U^\perp$ is space-like and V has the direct decomposition

$$V = U \oplus U^\perp,\ U \subset V_-,\ U^\perp \subset V_+. \qquad (\#)$$

In addition, for any decomposition of the type (#), one has $dim\ U = i$ [16.17].

Due to a counter-example, Noll assumed further that $ind\ V < \infty$. As a corollary to Theorem 9.1, he deduced that $V = Q(U) \oplus Q(U^\perp)$ for an orthogonal linear transformation $Q : V \to V$ [16.18]. What can be said now about the dimensionality of the signal subspaces of V? The answer gave the following of Noll's theorems:

Theorem 16.2. If $dim\ V - i \geq i$, then the maximal dimension of the signal subspaces of V is also given by i [16.19].

For a special case $ind\ V = 1$, Noll pointed out and proved four valuable results:

Theorem 16.3. $\{l^2 = -1, l \in V\} \Rightarrow \{v = \xi l + w, wl = 0, w \in V_+\}$ [16.20].

Theorem 16.4. If a vector is orthogonal to a non-zero time-like vector then it must be space-like [16.21].

Theorem 16.5. $\{u, v \in V_-\} \Rightarrow \{(u \cdot v)^2 \geq u^2 \cdot v^2$, and $(u \cdot v)^2 = u^2 \cdot v^2 \Leftrightarrow u, v$ are linearly independent$\}$ [16.22].

Theorem 16.6. $\{u, v, w \in V_-, u \neq 0, v \neq 0, w \neq 0\} \Rightarrow$ $\{(uv)(vw)(wu) < 0\}$ [16.23].

Noll introduced an equivalence relation $u \sim v$ for $u, v \in V_- = V_- \backslash \{0\}$, where 0 is the null vector: $u \sim v$ iff $u \cdot v < 0$. As it was easy to show, this relation defined a decomposition of V_- into two equivalence classes V_-^1 and V_-^2, to each of whom Noll adjoined the zero vector $\mathbf{0}$. Such V_-^1 and V_-^2 were called "directed time-cones". He proved two related results following from this definition:

Theorem 16.7. The time cone V_- is the union of two convex cones V_-^1 and V_-^2, which have only zero vector in common. Two time-like vectors u, v belong to the same cone if and only if $u \cdot v \leq 0$. If $u \in V_-$ belongs to one of the cones, then $-u$ belongs to the other, i.e., $V_-^1 = -V_-^2$.

Theorem 16.8. If $u, v \in V_-$ and $u \cdot v \leq 0$, then $\tau(u + v) \geq \tau(u) + \tau(v)$, and equality holds only when u and v are linearly dependent. $\tau(v) := \sqrt{-v^2}$, for any $v \in V_-$, and $\tau(v)$ is called the "duration of v" [16.24].

Thus, Noll constructed a *conical structure, corresponding to that of H. Minkowski*, and determined its properties. The next step was to introduce a *pseudo-Euclidean geometry*. Noll considered two more structures. One of them was a function $\sigma : E \times E \to \mathbf{R}$, where E was a set of points. The second structure was a group of automorphisms of E into itself, denoted by A, such that $\sigma(a(x), a(y)) = \sigma(x, y)$, $\forall x, y \in E$, $\forall a \in A$. Then, he defined a subgroup V of A, satisfying the following set of axioms:

A1: V is commutative.

A2: V is transitive.

A3: If $v \in V$ maps a point $x \in E$ onto itself, then v is the identity mapping.

A4: V is the underlying additive group of a real vector space and Φ ($\Phi : V \to \mathbf{R}$ and $\Phi(y - x) = \sigma(x, y)$) is a non-degenerate quadratic form on V; i.e., V can be given the structure of an inner product space such that

$$(y - x)^2 = (y - x)(y - x) = \sigma(x, y) .$$

The axioms were chosen with such care that the following *uniqueness theorem* held:

Theorem 16.9.   There is at most one subgroup V of the automorphism group A such that V satisfies the axioms **A1–A4**. If such a subgroup exists then its structure as an inner product space, as required for **A4**, is unique [16.25].

Now Noll could introduce a definition of the *pseudo-Euclidean geometry*: "A set E which is endowed with a structure defined by a real valued function σ on $E \times E$ is called a pseudo-Euclidean space if the axioms **A1–A4** are satisfied. The function σ will be called the separation function of E. The inner product space V determined by σ is called the translation space of E" [16.26]. As a corollary of Theorem 16.9, Noll deduced a generalization of the "well-known formula for rigid displacements" for the case of a pseudo-Euclidean space:

Theorem 16.10.   Let E be a pseudo-Euclidean space and q a point in E. Then every automorphism a of E has a unique representation of the form $a(x) = a(q) + Q(x - q)$, where Q is an orthogonal transformation of the translation space V [16.27].

Noll's next step was the introduction of certain mathematical concepts possessing a physical sense: (i) He called "events" the elements of E; (ii) He called "observers" the members of a family Ω of subsets of E which covered E; (iii) To each observer $\alpha \in \Omega$, he associated "a nonpositive separation function σ_α on α, which turned α into a one-dimensional pseudo-Euclidean space; (iv) He introduced a "signal relation" $\sim$ on E as a symmetric binary relation with the property that

$$\forall \alpha \in \Omega, \quad \forall x \in \alpha, \quad \exists y_1, y_2 \in \alpha : x \sim y_1 \wedge x \sim y_2 ;$$

he called "signal" a pair of events in the signal relation [16.28].
Walter Noll's axioms for Minkowskian chronometry were:

M1: $\forall x, y \in \alpha : \sigma(x, y) = \sigma_\alpha(x, y)$.

M2: (x, y) is a signal iff $\sigma(x, y) = 0$ iff $y \sim x$.

He called "Minkowskian domain" a set E with the following properties: it was endowed with a structure defined (i) by Ω, $\{\sigma_\alpha \mid \alpha \in \Omega\}$ and (ii) by the relation $\sim$, as described in (i)–(iv) and in the axioms **M1–M2**. For this Minkowskian domain he proved the following four results:

Theorem 16.11. There is at most one separation function σ which satisfies the axioms **M1** and **M2** [16.29].

Theorem 16.12. Every observer α is a time-like straight line in E: $\exists l \in V$, such that (i) $l^2 = -1$, and (ii) if $q \in a$, then $x \in \alpha$ iff $x = q + \xi l$, $\xi \in \mathbf{R}$ [16.30].

Theorem 16.13. Let α be an observer with direction vector l, let $q \in \alpha$, $x \in E$. An event $y = q + \eta l \in \alpha$ is then related to x by a signal iff η is a root of the equation

$$\eta^2 + 2l(x - q)\eta + (x - q)^2 = 0 \, ;$$

when $x \notin \alpha$, there are exactly two events

$$y_1 = q + \eta_1 l, \quad y_2 = q + \eta_2 l$$

in α that are related to x by a signal and the following equalities hold:

$$\sigma(q, x) = (x - q)^2 = -\eta_1 \eta_2,$$
$$(x - q)l = -(\eta_1 + \eta_2),$$
$$[l(x - q)]^2 + 4(x - q)^2 > 0 \quad [16.31] \, .$$

Theorem 16.14. The translation space V of a Minkowskian domain has index 1 [16.32].

After that Noll challenged the *problem of Suppes on temporal order* [16.33]. First of all, he exchanged the signal relation $\sim$ in (iv) for another one: (iv') "a binary relation $\rightarrow$ on E, such that

$$\forall \alpha \in \Omega, \ \forall x \in E \ \{\exists y_1 \in \alpha : y_1 \rightarrow x\} \ \{\exists y_2 \in \alpha : x \rightarrow y_2\};$$
$$x \rightarrow y \wedge y \rightarrow x$$

iff $x = y$". Noll called this relation a "directed signal relation". Obviously, $x \sim y$ iff $x \to y \lor y \to x$ [16.34]. In order to solve the problem of Suppes, it remained to add a third axiom to the axiomatical system of Minkowskian chronometry:

M3: If $\alpha \in \Omega$ and $y_1, y_2, z_1, z_2 \in \alpha$, and $y_1 \to p \to y_2$, $z_1 \to q \to z_2$ for some $p, q \in E$, then
$$\sigma_\alpha(y_1, z_1) + \sigma_\alpha(y_2, z_2) - \sigma_\alpha(y_1, z_2) - \sigma_\alpha(y_2, z_1) \geq 0 \,[16.35].$$

He applied the new directed signal relation to get a further definition of the *Minkowskian domain*: "A set E which is endowed with a structure defined by Ω, $\{\sigma_\alpha \mid \alpha \in \Omega\}$ and $\to$ as described under (i)–(iii) and (iv$'$) is called a directed Minkowskian domain iff the axioms **M1**–**M3** are satisfied" [16.36]. With the help of the directed signal relation, Noll defined another *relation for events, which he called "earlier"*: if $x \in E$, $y \in E$, then x was earlier than y, which was denoted $x < y$ iff $\exists p \in E$ such that $x \to p \to y$. For the new definition of Minkowskian domain, Noll proved two results:

Theorem 16.15. Every observer α has a unique direction vector l with the following property: for any two events $x, y \in E$
$$x < y \text{ iff } (y - x)^2 \leq l, (y - x) \cdot l \leq 0 \,[16.37].$$

Theorem 16.16. The relation $<$ is a partial order on E, which has the property that x and y are comparable iff $\sigma(x, y) \leq 0$ [16.38].

This relation $<$ allowed Noll to call the elements of the decomposition $V_- = V_-^1 \oplus V_-^2$ the *future time-cone* and the *past time-cone*.

According to John W. Schutz, the next axiomatic of Minkowski's chronometry was elaborated by George Szekeres, who presented it for publication in August 1966. In his paper one can find a critical evaluation of Noll's axiomatic. Szekeres called Noll's approach "traditional" and traced its roots back to Hermann Weyl's monograph "Space-Time-Matter" (1922). He described Noll's axiomatic as "much to be desired" from certain "metaphysical points of view". Szekeres wrote in particular: "In physical space-time, 3-space is essentially a space of simultaneity relative to a fixed observer and its properties can only be inferred indirectly, from kinematic observations relating to observers and light signals. It seems, therefore, inappropriate to build up kinematics from axioms which make use of assumed properties of 3-space, either explicitly or under the disguise of an inner-product space" [16.39].

In 1967, Mario Bunge proposed an axiomatic of special relativity, based not on the report of Minkowski, but on the works of Albert Einstein [16.40].

In 1973, Schutz published his axiomatical system for Minkowskian space-time in the form of a full-size monograph. He didn't evaluate Noll's contribution to the topic but restricted himself to a statement on the relation between the works of Suppes and Noll: their axiomatics were based "on the assumption of the invariance of the quadratic form $\Delta x_1^2 + \Delta x_2^2 + \Delta x_3^2 - c^2 t^2$ with respect to transformations between coordinate frames" [16.41].

Noll developed some ideas of his paper on Minkowskian chronometry in two jont publications with J.J. Schäffer, which appeared in 1976–1977. The first one contained three technical results for the future cones related to the Lorentz transformations [16.42]. The second paper dealt with the problem of analysis of order-isomorphisms between subsets of directed affine spaces, which concerned, in particular, the automorphisms of the causal-order structure in the Minkowskian chronometry [16.43].

In his fundamental treatise on finite-dimensional spaces, published in 1987, Noll collected materials related to his paper {2.24} on Minkowskian chronometry, in the §47 "Double-Signed Inner-Product Spaces". The theorems 5, 6 and 7, reformulated for V_+ instead of V_- , were offered by Noll there as exercises [16.44].

Noll's hope that "Minkowskian chronometry would become a branch of mathematics that was of interest purely for its aesthetic value" came to be true, but its new axiomatics ceased, as it seems, with the work of Schutz in 1973 [16.45].

Notes

1 In the Flatland

[1.1] E.A. Abbott, *Flatland: A Romance of Many Dimensions*, Princeton University Press, Princeton, 1991.

[1.2] According to Walter Noll's *Autobiographical Notes for 1925–1946*, his father was the originator of five patents.

[1.3] In 1918, Franz Noll was arrested and jailed for nine months for anti-war speaches.

[1.4] The house is still there today. Its address is the following: Fasanenstraße 32, Miersdorf by Zeuthen, Berlin.

[1.5] W. Noll, *Autobiographical Notes for 1925–1946*.

[1.6] W. Noll, *Autobiographical Notes for 1925–1946*.

[1.7] If the performance of a pupil at the primary school was modest, his parents had to pay the school fees for one year. During this year, the pupil had to obtain only good and very good marks in order to get a free place. If he didn't study well, the free place could be withdrawn after the year in favor of another candidate. Walter Noll's performance at the primary school was only average. Why could he then get a free place at once? One can find the answer in German history. In 1934, the last Jewish pupils were expelled from the high school for boys in Eichwalde. Their places became free and Walter Noll obtained one of them. As one of the teachers of the school remembers, there were no special critics of Jews in the class. The German Jews were generally considered as "super-intelligent" and "too waspish". The official anti-semitic propaganda didn't have much effect on the pupils because they considered it to be directed against abstract notions. Walter Noll didn't have Jewish relatives, and he could hope to get a university education which was his father's dream.

[1.8]   After 1933, the majority of German school teachers were forced to enter the Nazi association of teachers. Many of these "Nazi teachers", who only formally belonged to the Nazi party and were indiscriminately fired during the reeducation of Germany, committed suicide.

[1.9]   W. Noll, *Autobiographical Notes for 1925–1946.*

[1.10]   Ibid. Walter Noll said in an interview, published under the title "Walter Noll – The Language of Mathematics" in the newspaper "Focus", issued by the Faculty and Staff of Carnegie-Mellon University (Vol. 12, Nr. 2, October 1982): "There are a lot of mathematicians who hate numbers. I don't hate them, but I'm very bad at remembering numbers, I'm very bad at arithmetic. Most of what I do has very little to do with numbers, per se. I deal with structures that may or may not involve numbers, and if they do only very slightly. I tell my students: 'Mathematics is the art of avoiding unnecessary calculation'".

[1.11]   Paul Winkler remembered that Walter Noll didn't go in for sport at the TU in Berlin. In the USA, he learned to enjoy watching American sports. Resie Noll, his second wife, taught him snorkeling.

[1.12]   There was no religious education at the highschool for boys in Eichwalde, and Walter Noll's father was an atheist. It is because of this that Walter Noll is until today indifferent to religion.

[1.13]   A modern edition is: E.A. Abbott, *Flatland: A Romance of Many Dimensions*, Princeton University Press, Princeton, 1991. Abbott's success with the book laid partly in the demonstration of difficulties to prove the existence of a "land" of one more dimension by "measuring" the latter or indicating in what direction this extra dimension extended. Abbott introduced in a simple and attractive form how to do this abstractly, using the method of mathematical induction.

[1.14]   W. Noll, *Autobiographical Notes for 1925–1946.*

[1.15]   P. Karlson, *Du und die Natur: Eine moderne Physik für Jedermann*, Ullstein, Berlin, 1934.

[1.16]   W. Noll, *Autobiographical Notes for 1925–1946.*

[1.17] Two examples of this false understanding of mathematics presented by Rudolf Hohensee:

(i) given a_k, a_l, let's write an equality: $a_k = a_l$, then a can be crossed on the both sides of the equality, and we get: $k = l$, an absurd result;

(ii) $\{\sin^2 a + \cos^2 a = 1\} \Rightarrow \{a(\sin^2 + \cos^2) = 1\}$
$\Rightarrow \{a = 1/(sin^2 + cos^2)\}$.

[1.18] A copy of Walter Noll's notebook of physics is in the possession of the author. The original is presently kept by Rudolf Hohensee.

[1.19] Walter Noll exaggerated in his *Short Autobiography, Written in the Spring of 1949*, when he wrote: "From this time (4th grade) on, I always got very good marks in these subjects (mathematics and physics)". It took some time, at least until the winter semester of 1938/39, for Rudolf Hohensee to believe in Walter Noll's talent for mathematics and physics and to evaluate it properly.

[1.20] W. Noll, *Autobiographical Notes for 1925–1946*.

[1.21] In 1934, Rudolf Hohensee was a severely disciplined teacher. By 1941 he was no longer so and allowed his pupils to have parties in the school classroom, at which they danced and ate cakes. He often received telephone calls from their parents asking why their children were not at home at 9 p.m. Rudolf Hohensee was mobilized into the German army on June 19, 1941, shortly before the German attack on the Soviet Union. He was drilled and learned to stand and to run. He didn't become an officer and, as a mathematician, dealt with codes and secret writing. In the September of 1944, he became a prisoner of war of the Russians. His captivity lasted until 1947. As he returned to the Soviet sector of Berlin, he was able to resume teaching until 1950, when he was invited to face a communist commission regarding his political views. He didn't wait to be arrested and escaped to the US-sector. His wife, who lived there, tried to get a job for him at the TU. However, her chief, Professor Werner Schmeidler, a German, refused to help the refugee. It was a Hungarian, Istvan Szabó, who agreed to help and found him a position at a French school.

[1.22] Like many other pupils of the senior classes, Walter Noll had to stand night guard on the school roof when British and

American bombers came to bomb the innocent civilian population of Berlin.

[1.23]   W. Noll, *Autobiographical Notes for 1925–1946*.

[1.24]   One of Walter Noll's uncles, despite official prohibition, secretly kept in his house some books of Jews and German emigrants. Such books could also be bought unofficially on the "black market".

[1.25]   W. Noll, *Autobiographical Notes for 1925–1946*.

[1.26]   There are two good books about the "Reichsarbeitsdienst" and its creator: (i) Wiebke Stelling, Wolfram Mallebrein, *Männer und Maiden: Leben und Wirken im Reichsarbeitsdienst*, Preussisch-oldendorf, Verlag K.W. Schütz, 1979; (ii) Wolfram Mallebrein, *Konstantin Hierl: Schöpfer und Gestalter des Reichsarbeitsdienstes*, Hannover, National-Verlag, 1971.

[1.27]   W. Noll, *Autobiographical Notes for 1925–1946*.

[1.28]   In the *Autobiographical Notes for 1925–1946*, Noll called the course of E. Schmidt "Differential and Integral Calculus" or "Advanced Calculus". We use here the course title from the *Short Autobiography of Walter Noll, Written in Spring of 1949*. The following book was mentioned: B.L. van der Waerden, *Moderne Algebra*, Bd. 1–2, Springer, Berlin, 1930–1931.

[1.29]   K. Knopp, *Theorie und Anwendung der unendlichen Reihen*, Springer, Berlin, 1931.

[1.30]   Rudolf Hohensee called the German army the "Todes-schwadron" (squadron of death), because it suffered the highest level of losses in wartime.

[1.31]   *Handbuch zur deutschen Militärgeschichte 1648–1939*, Bd. IV, Abschnitt VII "Wehrmacht und National-sozialismus" (1933–1939), Bernard & Graefe Verlag, München, 1979, p. 514.

[1.32]   W. Noll, *Autobiographical Notes for 1925–1946*.

[1.33]   Walter Noll didn't want to go there.

[1.34]   W. Noll, *Autobiographical Notes for 1925–1946*.

[1.35]   W. Noll, *Short Autobiography, Written in Spring of 1949*.

[1.36]   H.B. Wells, *Being Lucky: Reminiscences and Reflections*, Indiana University Press, Bloomington, 1980, p. 303.

[1.37]   Letter of Walter Noll to Istvan Szabó, dated March 27, 1956.

[1.38] This room had the following address: Kaiserdamm Str. 27, Berlin-Charlottenburg.

[1.39] W. Noll, *Autobiographical Notes for 1946–1953*.

[1.40] W. Noll, *Autobiographical Notes for 1946–1953*. Istvan Szabó was one of the first teachers at the Technical University to recognize Walter Noll's mathematical talent. Szabó wrote, for example, in the "Zeugnis" of Walter Noll, dated August 3, 1955: "A decisive factor in his being taken on was that he (Walter Noll) had already impressed me as a young student due to his excellent performance of my lectures. His ability made him stand out from the crowd".

[1.41] W. Noll, *Autobiographical Notes for 1946–1953*. According to recollections of Paul Winkler, Walter Noll studied the following mathematical books thoroughly at that time: (i) K. Knopp, *Theorie und Anwendungen der unendlichen Reihen*, Springer-Verlag, Berlin, 1947; (ii) O. Perron, *Algebra I und II*, W. de Gruyter Verlag, Berlin, 1933; (iii) F. Tricomi, M. Krafft, *Elliptische Funktionen*, Akademische Verlagsgesellschaft, Leipzig, 1948.

[1.42] As Walter Noll wrote in his *Autobiographical Notes for 1925–1946*, the difference was that the Technical University offered an extra course of technical mechanics.

[1.43] In the *Autobiographical Notes for 1946–1953*, Noll remembered that his group was composed of about 40% French, 40% German and 20% other students.

[1.44] H.B. Wells, *Being Lucky: Reminiscences and Reflections*, Indiana University Press, Bloomington 1980, p. 305.

[1.45] W. Noll, *Autobiographical Notes for 1946–1953*.

[1.46] P. Fraenkel, *Notes on Discussion Group in Berlin (June 10, 1991)*, [Ms., unpublished.]

[1.47] According to the *Short Autobiography, Written in Spring of 1949*, Walter Noll visited the following courses at the Humboldt University in Berlin: (i) "Algebra I", "Algebra II", and "Eliminationstheorie", lectured by H.L. Schmid; (ii) "Zahlentheorie", lectured by H. Hasse; (iii) "Funktionentheorie II: Elliptische Funktionen", lectured by E. Schmidt; (iv) "Topologische Gruppen", lectured by H.F.A. Grell; (v) "Theoretische Logik II", lectured by K. Schröder. He also took part in the following two seminars there: (i) "Algebra",

conducted by H. Hasse and H.L. Schmid; (ii) "Laplace Transformations", conducted by E. Schmidt and K. Schröder.

[1.48]   This sum was sufficiently high since Walter Noll could save
from it enough money to buy all published volumes of "Elements
of Mathematics" by N. Bourbaki in July 1950 and to pay his
sightseeing trips.

[1.49]   In the *Autobiographical Notes for 1946–1953*, Walter Noll
wrote: "It was in Paris that I had my first contact with the work of
N. Bourbaki, which is a pseudonym that was used by a group of
brilliant French mathematicians including J. Dieudonné, A. Weil,
and H. Cartan. They were trying to reorganize, systematize, and
unify much of the mathematical knowledge known at that time".

[1.50]   In his letter to Y.A. Ignatieff, dated June 10, 1991, Ralph
Raimi remembered Walter Noll in Paris 1949–1950: "Walter was
a hard worker and very very serious. He had a great, maybe
exaggerated, respect for 'pure' mathematics as opposed to the
applied, or calculational, or algorithmic things that had apparently
been the substance of what he had been learning in Berlin".

[1.51]   Letter of Ralph Raimi to Y.A. Ignatieff, dated June 10,
1991.

[1.52]   Ibid.

[1.53]   Ibid.

[1.54]   Ibid.

[1.55]   Ralph Raimi maintained that Walter Noll could already
have had the intention to emigrate to the USA in 1949–1950.

[1.56]   It was a remarkable result. In his *Autobiographical Notes
for 1946–1953*, Noll remembered: "It was not easy to pass. For
example, only 49 out of 296 candidates passed the examination in
'Calcul Différentiel et Integral', when I took it; of those 49, only
four got a 'très bien' (A), and another four a 'bien' (B)".

[1.57]   Ralph Raimi remembered that "a French student, having
come from a lycée and starting from the beginning, would have
spread ... (examinations for the degree of Licencié ès Sciences) out
over three to four years".

[1.58]   The Noll family house in Berlin was sold in 1950.

[1.59]   In the "Zeugnis" of Noll's diploma examinations, dated
May 5, 1951, one can find the following subjects and notes: (i) *pure*

mathematics: "very good"; (ii) *applied mathematics*: "very good"; (iii) *theoretical physics*: "good"; (iv) *theory of functions*: "very good"; (v) *group theory*: "good"; (vi) *higher technical mechanics*: "very good".

[1.60]   See the *Autobiograpical Notes of Walter Noll for 1946–1953*, where he gave this translation from German. Walter Noll's idea about fundamental structures relates naturally to the "classical" part of mathematics.

[1.61]   Walter Noll and Klaus André lodged in the same flat by Frau Emilie Neumark in Berlin-Wilmersdorf between October 1951 and August 1953. They spent much time together playing a chinese board-game "GO", visiting movies, and listening to the radio. After graduating from the Free University, André received a two-year contract to assist the famous German mechanician Georg Hamel to write a textbook on continuum mechanics. However, this work was interrupted by Hamel's death. This textbook was finished by Istvan Szabó. It appeared under the following title: G. Hamel, *Mechanik der Kontinua*, Teubner, Stuttgart, 1956. Before leaving Germany in September of 1955, Noll arranged that André received a scientific assistantship with Szabó. It was André who gave Szabó the idea to offer Noll a vacant professor position at the Technical University. In 1959, André got a Ph.D., and Dietrich Morgenstern was a scientific advisor for his dissertation. André later worked at the famous Hahn-Meitner Institute for Reactor Research Technics. Since 1984, he has been the director of the Department for Scientific-Technical Services at the Konrad-Zuse Center for Information Technics in Berlin.

[1.62]   The friendship of Walter Noll and Paul Winkler was different to that with Klaus André. Winkler was a pure mathematician and worked on a dissertation on number theory. When Noll wanted to relax he spent time with André. If he needed to open his soul he went to Winkler. The latter knew the character of Noll as nobody else. On March 3, 1991, Winkler wrote to Y.A. Ignatieff: "Walter Noll was always open and direct. He spoke his mind and did not beat about the bush. He seems to be pitiless to the mistakes of others whether these were his colleagues, other students, or – at that time – one of his professors. Conniving, even when the end justified the means, was not something he was noted for". Following his assistantship,

Paul Winkler went to the insurance business. He became interested in computers and worked for twelve years at IBM in Germany. The last seventeen years of his career, Paul Winkler spent in a responsible position in the German State Ministry of Labour and Social Security.

[1.63]   The friendship between Walter Noll and Dietrich Morgenstern was also a special kind. Noll respected him as a fair counterpart in the competition to be the best mathematical student at the Technical University. When Noll emigrated to the USA in 1955, Morgenstern became a dozent at the Berlin Technical University. However, Noll received a position of full professor two years earlier than his counterpart. Morgenstern was a professor at the Universities of Münster, Freiburg, and Hannover. Since 1986, he has been a professor emeritus of mathematical stochastics at the University of Hannover.

2 Slave of Istvan Szabó

[2.1]   Ernst Mohr was a good teacher and an original mathematician. His arrest by the German secret police in 1944 as an opponent of the Nazi regime put Mohr very high in Walter Noll's estimation.

[2.2]   There were at least two editions of this handbook: (i) E.F. Beckenbach and R. Bellmann, *Inequalities*, Springer-Verlag, Berlin, 1961, p. 46 (§46: "A Result of Mohr and Noll"); (ii) E.F. Beckenbach and R. Bellmann, *Inequalities*, 2. Edition, Springer-Verlag, Berlin, 1965, p. 46 (§46: "A Result of Mohr and Noll").

[2.3]   In 1951–1955, Walter Noll had a room in the flat of Frau Emilie Neumark at the following address: Binger Str. 10, Berlin-Wilmersdorf.

[2.4]   W. Noll, *Autobiographical Notes for 1946–1953*.

[2.5]   Ibid.

[2.6]   See the "Zeugnis" of Walter Noll, written by Istvan Szabó and dated August 3, 1955. The following books are mentioned: (i) I. Szabó, *Mathematische Formeln und Tafeln, Hütte's Des Ingenieurs Taschenbuch*, Akademischer Verein Hütte e.V., Berlin, 1955;

(ii) I. Szabó, *Integration und Reihenentwicklungen im Komplexen, Gewöhnliche und Partielle Differentialgleichungen, R. Rothe's Höhere Mathematik für Mathematiker, Physiker und Ingenieure*, Teil VI, B.G. Teubner, Leipzig, 1953.

[2.7] See the "Zeugnis" of Walter Noll, written by Istvan Szabó and dated August 3, 1955. The following book is mentioned: I. Szabó, *Technische Mechanik*, Springer-Verlag, Berlin, 1956.

[2.8] W. Noll, *Autobiographical Notes for 1953–1956*. The reasons for Noll's emigration to the USA can also be found in his interview for the paper "Focus" (Carnegie-Mellon University, Vol. 12, Nr. 2, October 1982), where he said in particular: "There are two reasons why the United States ... became a major factor in mathematics: one was all the refugees that came here; the other was that during and after the war the federal government pumped a lot of money into it and attracted people. I'm an example, and there were many people like me. In Europe, you couldn't get a job – it was burned out; it was easier to operate here". As in many other cases, Walter Noll exaggerates here and doesn't call things with their proper names.

[2.9] Ibid. Istvan Szabó wrote in the "Zeugnis" of Walter Noll: "After his final examination, where he received a Diploma in Engineering 'with honors', he became a scientific assistant in my department. Through this activity Dr. Noll was able to apply his mathematical knowledge to the problems of physics and especially those of mechanics. Here, Dr. Noll proved himself very good in scientific and pedagogical respects and during this time was my best and most valuable support in all difficult problems, especially mathematical ones".

3 Truesdell

[3.1] See the preface, written by C.A. Truesdell, for the following book: *The Foundations of Mechanics and Thermodynamics: Selected Papers of W. Noll*, Springer-Verlag, Berlin, 1974.

[3.2] Ibid.

[3.3] W. Noll, *Autobiographical Notes for 1946–1953*.

[3.4]   Letter of C.A. Truesdell to G. Hamel, dated August 13, 1954.

[3.5]   Ibid. In the private archive of Truesdell in Baltimore, there is the following list of subjects, offered to Walter Noll at the preliminary doctorate examination in 1953:

(i) *Analysis*: theory of functions of a complex variable; conformal mapping; elliptic functions; ordinary differential equations in the real domain; ordinary differential equations in the complex domain and special functions (hypergeometric, Bessel's, Mathieu's functions); partial differential equations of first and second order; potential theory; integral equations; calculus of variations; Laplace and Fourier transforms; vector and tensor analysis; theory of measure and integration; Banach- and Hilbertspace; functional analysis; kernel functions; Lie groups; theory of distributions (L. Schwartz);

(ii) *Algebra*: linear algebra (vector-spaces, matrices, tensors etc.); higher algebra (groups, fields, rings, Galois theory, representations, algebras); topological groups; theory of numbers (valuations, algebraic number fields);

(iii) *Geometry and Topology*: differential geometry and Riemannian geometry; projective geometry; point-set topology; algebraic topology; non-Euclidean geometry;

(iv) *Mechanics and Physics*: technical mechanics; theoretical mechanics; theory of elasticity; electrodynamics; thermodynamics; optics; quantum mechanics;

(v) *Unclassified*: set-theory; symbolic logics; numerical mathematics; theory of games.

In the *Autobiographical Notes for 1953–1956*, Noll remembered: "I had to take the qualifying examination without much preparation, but I passed despite the fact that I could not answer several questions, perhaps because the examiners realized that I could think on my feet".

[3.6]   Dietrich Morgenstern made an important contribution to early American rational mechanics. Unfortunately, it hasn't been recognized properly in scientific literature.

[3.7]   Letter of C.A. Truesdell to G. Hamel, dated August 13, 1954.

[3.8]   This belief can be traced back to the school years of Walter Noll and to his reading of the book "Du und die Natur" by P. Karlson. Noll wrote in his *Autobiographical Notes for 1925–1946*: "I started to read a book entitled 'Du und die Natur', a popularization of modern physics. I was fascinated by this book; I learned about atoms and molecules, electrons, protons, and neutrons, the theory of relativity and quantum mechanics". This belief turned with years into a dogma for Noll. For example, he wrote in the paper "The Foundations of Classical Mechanics in the Light of Recent Advances in Continuum Mechanics" (1959): "It is known that continuous matter is really made up of elementary particles. The basic laws governing the elementary particles are those of quantum mechanics. The science that provides the link between these basic laws and the laws describing the behavior of gross matter is statistical mechanics" (p. 266).

[3.9] In his letter to G. Hamel, dated August 13, 1954, Clifford Truesdell remarked about this choice: "I believe it deserves special mention that Mr. Noll is highly competent in modern pure analysis, including the developments of the French School, and is highly regarded by the analysts here. However, instead of writing in analysis, which would have been easy for him, he chose an entirely new subject (continuum mechanics)".

[3.10] It is interesting that Clifford Truesdell agreed to be the advisor for Noll's thesis. He knew, of course, that Noll's knowledge of the subject was very limited. We find an answer in the letter of Truesdell to G.A. Oravas, dated January 21, 1969: "I asked Hamel in 1952 to recommend good men. That is why Noll and Morgenstern came to Indiana then. Apart from myself, they are the only ones among all those I have named whose basic training and initial interest was in pure mathematics, but both of them were also Szabó's assistants in Berlin, where they taught engineers and draughted large parts of engineering textbooks since published under the names of professors there. The common accusation that modern work on the foundations of mechanics is done by pure mathematicians unfamiliar with real problems is obvious nonsense to anyone who knows who the people are and where they come from".

[3.11] W. Noll, *Autobiographical Notes for 1953–1956*.

[3.12] Ibid.

[3.13] Letter of C.A. Truesdell to G. Hamel, dated August 13, 1954.

[3.14] W. Noll, *Autobiographical Notes for 1953–1956*.

[3.15] The article was reprinted in the following book: *International Science Review Series*, Vol. 8/II, Gordon & Breach, New York 1965.

[3.16] This announcement is available in the private archive of C.A. Truesdell in Baltimore.

[3.17] Letter of C.A. Truesdell to G. Hamel, dated August 13, 1954.

[3.18] In 1974, Clifford Truesdell excluded Noll's thesis from the volume of his selected works with the following comment: "Had influence already manifest been the basis of the choice, it would have dictated inclusion of No. 2, which was Noll's thesis, but now that work is in part obsolete, in part available through the intermediacy of a dozen books and a hundred papers".

[3.19] See {2.2}, p. 5.

[3.20] The symmetry of S means nothing more than an introduction of the law of moment of momentum, discovered by Leonard Euler. One can find further details in the following paper: C.A. Truesdell, "A Program Toward Rediscovering the Rational Mechanics of the Age of Reason", *Archive for History of Exact Sciences* **1** (1960), 3–36.

[3.21] *The Oxford English Dictionary*, Vol. III, 2. Edition, Clarendon Press, Oxford, 1989, p. 791, s.v. "Constitutive Equation".

4 Noll's Odyssey from Los Angeles to Pittsburgh

[4.1] Walter Noll's first flat had the following address: 1108 $^1/_2$ Exposition Boulevard, Los Angeles 6.

[4.2] Walter Noll's second flat had the following address: 4010 Palmwood Drive, Los Angeles 8.

[4.3] The salary of Walter Noll as an instructor at the University of Southern California was US$ 4500 for the academic year. In the summer of 1956, he also received US$ 1500 under a U.S. Government research grant.

[4.4] According to Noll's *Autobiographical Notes for 1953–1956*, R. Finn voted against offering him a position at the University of Southern California, because Noll was a German.

[4.5] C.A. Truesdell and D. Gilbarg, Noll's teachers at Indiana University, nominated him for this appointment at the Carnegie Institute of Technology.

[4.6] See {2.4, p. 6}. This paper was just a good technical result to satisfy the "publish-or-perish" rule, common at US-universities.

[4.7] Istvan Szabó did his best to find a good position for Walter Noll in Berlin. He didn't believe that Noll's decision to leave Germany was final. Certainly, Szabó knew the content of Clifford Truesdell's letter to G. Hamel, dated August 13, 1954, where the American wrote: "He (Walter Noll) has received offers of positions here, but he prefers to return to Germany, where there will soon be an opening worthy of his talents".

[4.8] The problem of non-existence of plane Stokes flow belongs to "scientific absurdities in the impressive uniform of formulae and theorems" (see: J. Schwartz, "The Pernicious Influence of Mathematics on Science", *Proceedings of the 1960 International Congress on Logic, Methodology and Philosophy of Science*, Stanford University Press, Stanford, California, 1962, pp. 356–360). Surely, it was R. Finn who proposed the topic of this article. Walter Noll demonstrated there his brilliant mathematical technique.

5 "Walter Noll, Our Teacher"

[5.1] In 1912, the Carnegie Institute of Technology was chartered "to provide undergraduate education and to foster research and creative attainment" (see: *The Encyclopedia Americana*, Vol. 5, Grolier, Danburg, 1985, p. 684).

[5.2] Ibid.

[5.3] See: the brochure *Mellon College of Science*, p. 3, and *The World of Learning 1992*, 42. Edition, Europa Publications Ltd, London, 1991, pp. 1795–1797.

[5.4]   In the brochure *Mellon College of Science*, the Department of Mathematics was called "one of the outstanding national centers for applied mathematics" (pp. 1–2).

[5.5]   In this publication, R.J. Duffin stands as the first author. Evidently, the idea of this paper belongs to him. Noll's contribution to it was the use of the mathematical method, which had already proved its efficiency in the solution of the problem of Stokes flow ({2.7}, p. 191; {2.5}).

[5.6]   Letter of J.W.T. Youngs to W. Noll, dated December 3, 1958.

[5.7]   *Actes du 4ᵉ Colloque International de Logique et Philosophie des Sciences*, Gauthier-Villars Editeur, Paris, 1963, p. 6.

[5.8]   Ibid., p. 7.

[5.9]   Ibid., p. 56.

[5.10]   In a letter to B. Bernstein, dated January 16, 1960, Noll described the content of his lecture in the following words: "I should like to present some results which cast light on the manner in which Newtonian fluids approximate general fluids for slow motions".

[5.11]   Letter of R. Raimi to Walter Noll, dated September 15, 1960.

[5.12]   Letter of Walter Noll to J. Auslander, dated September 13, 1960.

[5.13]   In his letter to D.C. Leigh, dated April 6, 1962, Noll described the content of his lecture at Princeton University as follows: "I would attempt to present a review of some of the ideas that have been introduced in this field during the last few years, and I would try to explain some new results on the viscometry of simple fluids and the kind of kinematics connected with these results".

[5.14]   Walter Noll wrote to D.C. Leigh on May 18, 1962: "It is very gratifying to find that the circle of people who wish to join the 'Society of the Body B' (Truesdell's term) is enlarging".

[5.15]   The invitation for Noll's lecture at the U.S. Naval Research Laboratory was signed by Horace M. Trent. It was dated October 23, 1962.

[5.16]   Letter of E. Zaustinski to Walter Noll, dated March 2, 1964.

[5.17]   In the enclosed brochure, Noll read the following description: "The current seminar will be devoted to the foundations of physics and, in particular, to an analysis and criticism of the fundamental theories of physics from classical mechanics to quantum theory. The twelve invited lecturers and commentators, jointly with members of the physics and philosophy staffs of the University of Delaware, will discuss the meaning of axiomatic foundations of some of the basic physical theories, will point out some of the unresolved difficulties in foundations of some of the basic physical theories, and will point out some of the unresolved difficulties in foundational research".

[5.18]   Letter of Mario Bunge to Walter Noll, dated September 21, 1964.

[5.19]   Letter of Walter Noll to Mario Bunge, dated September 29, 1964.

[5.20]   Letter of Walter Noll to R. Plunkett, dated December 12, 1963.

[5.21]   In his letter to G. Oravas, dated January 21, 1969, C.A. Truesdell bitterly wrote about this collaboration: "Noll is a poor correspondent. Even when we were collaborating he would leave my urgent questions unanswered for months. . . . He is very friendly and kindly but extremely absent-minded".

[5.22]   See {2.2R}, {2.9R1}, {2.11R1}, {2.12R}, {2.14R1}, {2.15R1} and {2.17R1}.

[5.23]   Philipp Lenard wrote: "However, there is one thing that we must take special care to avoid: Namely, a sort of *self-torment* which we carry out in our own country purely for reasons of scruples. The German can seldom do well enough for another German" (see: P. Lenard, *England und Deutschland zur Zeit des großen Krieges*, Carl Winter, Heidelberg, 1914, p. 13).

[5.24]   In his letter to Walter Noll, dated September 16, 1966, D. Frederick wrote: "One of the highlights was your series of outstanding, informative lectures . . . Many of the participants have asked that I relay to you their appreciation of your work and presentations".

[5.25]   The publication of this monograph was arranged by B.D. Coleman, who became the Editor-in-Chief of the Series "Springer Tracts in Natural Philosophy" in 1967 after the resignation of C.A. Truesdell from this post.

[5.26] Letter of T.S. Chang to Walter Noll, dated November 29, 1967.

[5.27] Letter of A. Goldburg to Walter Noll, dated January 4, 1968.

[5.28] Letter of M. Kruscal to Walter Noll, dated March 14, 1968.

[5.29] Letter of Walter Noll to M. Kruscal, undated.

[5.30] Letter of G. Capriz to Walter Noll, dated May 31, 1967.

[5.31] In his famous book "The Thermodynamics of Simple Materials with Fading Memory" (1972), Professor W.A. Day ascribed to B.D. Coleman all the fame for the construction of a "rational and exact theory of thermodynamics" (p. 2). One can only speculate about the reasons for this mistake. However, Coleman is certainly responsible for allowing this mistake to flatter his ears.

[5.32] Walter Noll wrote in his letter: "Perhaps you permit me to propose that you invite my colleague, Professor B.D. Coleman. Actually, he might be more suitable than myself, for I have not worked on rheology for some time, while he just finished an interesting study of the theory of the behavior of bubbles in fluids with memory".

[5.33] Letter of D. Muster to Walter Noll, dated June 4, 1968.

[5.34] Letter of M.W. Wilcox to Walter Noll, dated August 28, 1968.

[5.35] See {2.9}, {2.31}, {2.32}.

[5.36] Letter of W. Noll to E. Enochs, dated March 23, 1969.

[5.37] Walter Noll's account of his SIAM-lecture at the Geneva College in Beaver Falls, PA, in 1970.

[5.38] Walter Noll's account of his SIAM-lecture at the Shippensburg State College in Shippensburg, PA, in 1970.

[5.39] Walter Noll's account of his SIAM-lecture at the Millersville State College in Millersville, PA, in 1970.

[5.40] Walter Noll's account of his SIAM-lecture at the Marquette University in Milwaukee, Wisconsin, in 1970.

[5.41] Walter Noll's account of his SIAM-lecture at the University of Wisconsin-Milwaukee in Milwaukee, Wisconsin, in 1970.

[5.42] Walter Noll's account of his SIAM-lecture at the Denison University in Granville, Ohio, in 1970.

[5.43] On September 17, 1970, W.E. Langlois offered Walter Noll to give lectures in Chicago, Cleveland, Madison, Urbana, South Bend and other nearby cities to Notre Dame. Then, according to his calculations, Noll could earn US$ 125. Obviously, this money didn't cover the loss of time of the great mathematician. One should be Walter Noll not to take this offer as an insult.

[5.44] This list of topics was enclosed in the SIAM Visiting Lectureship Program brochure for 1970–1971.

[5.45] Walter Noll's account of his SIAM-lecture at the University of Alabama in Huntsville, Alabama, in 1971.

[5.46] Walter Noll's account of his SIAM-lecture at the Chipola Junior College in Marianna, Florida, in 1971.

[5.47] Walter Noll's account of his SIAM-lecture at the Franklin and Marschall College in Lancaster, PA, in 1971.

[5.48] In a letter to Walter Noll, dated March 10, 1971, John M. Petrie wrote: "I wish to assure you that your efforts did not go unnoticed or unappreciated. I'm sure that your introduction to some of the concepts of relativistic time will serve to encourage many of our students to delve further into the subject and thus, I feel certain that your lectures were a valuable service to the academic community".

[5.49] Walter Noll's account of his SIAM-lecture at the Villanova University in Villanova, PA, in 1971.

[5.50] Walter Noll's account of his SIAM-lecture at the University of Arizona in Tucson, Arizona, in 1971.

[5.51] Walter Noll's account of his SIAM-lecture at the University of Colorado in Boulder, Colorado, in 1971.

[5.52] Walter Noll's account of his SIAM-lecture at the University of Toledo in Toledo, Ohio, in 1971.

[5.53] This publication ended a scientific collaboration between Walter Noll and H.D. Dombrowski. Noll never returned to this paper in his further publications.

[5.54] Letter of R. Schramm to Walter Noll, dated May 29, 1972.

[5.55] Letter of B. Lauvere to Walter Noll, dated December 10, 1975.

[5.56] Letter of Walter Noll to S. Goldman, dated October 28, 1984.

[5.57] Letter of B. Angeloni to Walter Noll, dated March 26, 1985.

[5.58] Letter of Enrico Magenes to Walter Noll, dated January 12, 1989.

[5.59] W. Noll, *My Trip to the Soviet Union, Nov. 13–25, 1989.*
[5.60] Ibid. Walter Noll wrote there: "The organization of the meeting was one of the worst I have ever seen. There was no program, no time-table, no list of speakers or topics of lectures. Everything went by word of mouth, and half the time I was not sure what was going on. However, the organizers meant well and their hospitality was overhelming".

6 Mathematics Educator, Researcher and Professor, Walter Noll

[6.1] See two documents: (i) W. Noll, *The Mathematical Studies Program: Remarks by Walter Noll*; (ii) "Walter Noll – The Language of Mathematics", an interview of Walter Noll for the "Focus", a publication of the Faculty and Staff of Carnegie-Mellon University, Vol. 12, Nr. 2, October 1982.
[6.2] "Walter Noll – The Language of Mathematics", an interview by Walter Noll for the "Focus", a publication of the Faculty and Staff of Carnegie-Mellon University, Vol. 12, Nr. 2, October 1982.
[6.3] Ibid.
[6.4] W. Noll, *The Mathematical Studies Program: Remarks by Walter Noll.*
[6.5] Ibid.
[6.6] Ibid.
[6.7] W. Noll, *The Decline of the Professor.*
[6.8] W. Noll, *The Mathematical Studies Program: Remarks by Walter Noll.*
[6.9] Ibid.
[6.10] W. Noll, *Courses Taught by Walter Noll at CMU 1956–1993.*
[6.11] Ibid.
[6.12] W. Noll, *The Decline of the Professor.*
[6.13] The Annual Report for 1978 by Walter Noll at the Carnegie-Mellon University.

[6.14]   The Annual Report for the Fall Semester of 1981 and the Spring Semester of 1982 by Walter Noll at the Carnegie-Mellon University.

[6.15]   The Annual Report for the Fall Semester of 1983 and the Spring Semester of 1984 by Walter Noll at the Carnegie-Mellon University.

[6.16]   The Annual Report for the Fall Semester of 1985 and the Spring Semester of 1986 by Walter Noll at the Carnegie-Mellon University.

[6.17]   The Annual Report for the Fall Semester of 1986 and the Spring Semester of 1987 by Walter Noll at the Carnegie-Mellon University.

[6.18]   The Annual Report for the Fall Semester of 1987 and the Spring Semester of 1988 by Walter Noll at the Carnegie-Mellon University.

[6.19]   The Annual Report for the Fall Semester of 1988 and the Spring Semester of 1989 by Walter Noll at the Carnegie-Mellon University.

[6.20]   The Annual Report for the Fall Semester of 1989 and the Spring Semester of 1990 by Walter Noll at the Carnegie-Mellon University.

[6.21]   W. Noll, *Research Plans* ($\sim$1989).

[6.22]   W. Noll, *The Decline of the Professor*.

[6.23]   Ibid.

[6.24]   Ibid.

[6.25]   Ibid.

[6.26]   These two publications by W. Van Buren are: (i) "On the Existence and Uniqueness of Solutions to Boundary Value Problems in Finite Elasticity", *Research Report* 68-ID7-MEKMA-RI, Proprietary Class 3, July 31, 1968, Mechanics Department of the Westinghouse Research Laboratories, Pittsburgh, PA, 15235; (ii) "On the Traction Boundary Value Problem of Infinitesimal Elastic Deformations Superimposed on a Fixed Large Deformation", *Research Report* 67-ID7-ELPLA-R2, Proprietary Class 3, July 26, 1967, Westinghouse Research Laboratories, Pittsburgh, PA, 15235.

[6.27]   An inter-office correspondence of Carnegie-Mellon University, addressed to the university community by Richard M. Cyert on

September 11, 1981, and concerned with the freedom of the members of the community.

[6.28] L.E. Bragg, "On Relativistic Worldlines and Motions and on Non-Sentient Response", *Archive for Rational Mechanics and Analysis* 18/2 (1965), 127–166.

[6.29] An inter-office correspondence of the Carnegie-Mellon University on the subject "Student-Alumni Associates", dated February 8, 1974, and signed by Ronald Broughman on the side of the Alumni Association.

[6.30] Walter Noll resigned as a member of the Editorial Board of the Archive for Rational Mechanics and Analysis in 1976 after Helga Noll's death.

[6.31] In a letter to Noll, dated February 19, 1973, B.D. Coleman wrote: "I want to express you my sincere thanks for your efforts as a co-editor of this series [Springer Tracts in Natural Philosophy]". The series was finished at the Springer-Verlag in about 1978. Klaus Peters wrote in a letter to Noll, dated March 12, 1973: "We have come to the conclusion to discontinue the series as the projects which have already been contracted are completed". According to the Springer publication "Books Written or Edited by C. Truesdell", the last (29th) volume of the series was "Relativistic Theories of Materials" by A. Bressan in 1978.

[6.32] Letter of Walter Noll to Harding Bliss, the Editor of A.I.Ch.E. Journal, dated June 20, 1966. On May 3, 1976, Noll wrote to the famous mathematician, Morton Hamermesh: "The paper [to be reviewed] deals with continuous media with microstructure, a subject I know nothing about and do not want to know anything about".

[6.33] Walter Noll wrote in a letter to H.M. Srivastava on March 1, 1978: "In my opinion, the paper should not be published. . . . I find it very unlikely that the constitutive equation . . . used by the author, applies to any real material, because it is obtained by arbitrary assuming that a certain material constant is zero. . . . I doubt that the mathematical analysis, which is routine though tedious, sheds light on any real physical phenomenon".

[6.34] According to "Who's Who in America", Noll is also a member of the third professional society: the Mathematical Association of America. From an undated letter of Linton E. Grinder, the national president of the Scientific Research Society of North America (Sigma XI), to the Graduate School of the University of Florida in Gainesville, Florida, one can conclude that Noll also had some connection with this society in about 1974.

7 Algebra, Geometry, and Analysis

[7.1]   See {2.33R}, p. 1.

[7.2]   See {2.36}. According to the Science Citation Index, this paper was to appear in: *Proceedings of the American Mathematical Society* **23** (1969), 1–69.

[7.3]   Letter of H.D. Dombrowski to Y.A. Ignatieff, dated December 11, 1991. Dombrowski's habilitation work was published in Göttingen in the *Nachrichten der Akademie der Wissenschaften*, mathematisch-physikalische Klasse (Vandenhoeck and Ruprecht), 1966, pp. 19–43.

[7.4]   See this dissertation in: *Mathematische Annalen* **149** (1963), 71–96. Dombrowski wrote on this topic: "In this work two-dimensional representation modules of the elliptic modular group Γ and its main congruence group $\Gamma(2)$ will be studied. We are only interested in the representation modules of analytical functions on the upper half-plane" (p. 71). His results received a high evaluation in: H.S. Holdgrün, "Irreduzibilität und Äquivalenz von Darstellungen", *Mathematische Zeitschrift* **96** (1967), 343–354.

[7.5]   See the following papers: (i) H.D. Dombrowski, K. Horneffer, "Der Begriff des physikalischen Systems in mathematischer Sicht", *Nachrichten der Akademie der Wissenschaften in Göttingen*, mathematisch-physikalische Klasse (1964), 67–100; (ii) H.D. Dombrowski, K. Horneffer, "Die Differentialgeometrie des Galileischen Relativitätsprinzips", *Mathematische Zeitschrift* **86** (1964), 291–311; the authors wrote on its content: "In this discourse, the Galilean principle of relativity will be applied consistently in Newtonian mechanics; one arrives in this way at the concept of the Galilean man-

ifold. The differential geometry of Galilean manifolds is the mathematical subject of the present work" (p. 291).

[7.6] H.D. Dombrowski, K. Horneffer, "Das verallgemeinerte
Galileische Relativitätsprinzip", *Nachrichten der Akademie der Wissenschaften in Göttingen*, mathematisch-physikalische Klasse
(1964), 233–246.

[7.7] Letter of H.D. Dombrowski to Y.A Ignatieff, dated December 11, 1991. Dombrowski's report at the Oberwolfach Mathematical Research Institute was entitled "Über die Invarianz der Naturge-
setze", and it was not published in the Conference Proceedings. One
can get an impression of its content and evaluate its influence upon
Walter Noll from the following citation from it: "As is known, the
role of invariant properties in physics is judged in many different
ways. In the following, I consider only the so-called geometrical invariance and do not differentiate between invariance and covariance.
From the wide spectrum of opinions, two extreme positions will be
mentioned briefly: a) General covariance is necessary for epistemo-
logical reasons (Einstein); b) General covariance is achievable in
every theory for physical and mathematical reasons (Kretschmann).
These two hypotheses have opposed each other irreconcilably for
50 years. In view of the great importance of invariance principles
in physics, a clarification of the situation is a worthwhile purpose.
In the first part of my lecture, I will analyse the concept of invariance and some examples; in the second I will say a few words about
the foundation of the invariance of the *laws of short-range effects*"
(p. 1).

[7.8] Ibid.

[7.9] Ibid.

[7.10] Walter Noll lived far away from the university. He drove
there by car when he had lectures, office hours or other university
duties. However, he used to do his creative mathematical work at
home, where Dombrowski couldn't come too often due to American
standards of life.

[7.11] Letter of H.D. Dombrowski to Y.A. Ignatieff, dated December 11, 1991.

[7.12] Ibid.

[7.13] H.D. Dombrowski, "On Simultaneous Measurements of Incompatible Observables", *Archive for Rational Mechanics and Analysis* **35** (1969), 178–210.

[7.14] In {2.36}, Walter Noll's name occupies second place in the list of authors after H.D. Dombrowski.

[7.15] See {2.36}, p. 209.

[7.16] Ibid., p. 213.

[7.17] Ibid., pp. 207, 276–278.

8 Finite-Dimensional Spaces

[8.1] Letter of W. Noll to Y.A. Ignatieff, dated December 21, 1994.

[8.2] See {1.1}, where one can read on the front page the following lines: "Notes prepared by C.C. Wang on the basis of a course given by Professor W. Noll at Johns Hopkins, 1962–1963".

[8.3] Ibid., p. 86.

[8.4] Ibid., pp. 86–89.

[8.5] See {1.2}, pp. 20–35.

[8.6] See the preface to {1.3}.

[8.7] The version of Noll's notes on tensor analysis, included in {1.3}, is very short in comparison with the original Baltimore version: only 16 pages.

[8.8] See {1.3}, pp. 93–108.

[8.9] C.A. Truesdell, *A First Course in Rational Continuum Mechanics*, Vol. 1, 2. Edition, Academic Press, Boston, 1991, pp. 301–341. Truesdell wrote there regarding the appendix with mathematical tools of continuum mechanics: "The material listed here is drawn largely from unpublished notes leading to W. Noll's 'Finite-Dimensional Spaces', Volume 1, 'Algebra, Geometry, and Analysis', Dordrecht etc., Martinus Nijhoff, 1987" (p. 301).

[8.10] See {1.5}, p. IX.

[8.11] Ibid., p. VII.

[8.12] In {1.5}, Walter Noll stressed the role of J.J. Schäffer as a collaborator, who took part in the development of the system of notation and terminology used in the treatise.

[8.13] These materials are in the possession of the author. In {2.48}, Walter Noll and E.G. Virga remarked that a preliminary version of Chapter 3 of the second volume of Noll's treatise "Finite-Dimensional Spaces: Algebra, Geometry, and Analysis" served as the basis for Section 3 of the following paper: M.E. Gurtin, A.I. Murdoch. "A Continuum Theory of Elastic Material Surfaces", *Archive for Rational Mechanics and Analysis* **57** (1974), 291–323.

9 Foundations of Rational Continuum Mechanics

[9.1] Walter Noll's mathematical works show that he tried to solve the sixth problem of David Hilbert: to build axiomatical systems at different levels of the physical organization of matter and to secure connections among them. This conclusion is supported by the data in Table N.1:

Table N.1. Walter Noll's works on the axiomatization of physics

Physical discipline	Begin	Publications
Quantum Mechanics (QM)	1977	{2.42}, {2.43} etc.
Transition from QM to SM	–	–
Statistical Mechanics (SM)	1970	{2.35} etc.
Transition from SM to CM	1955	{2.3}
Continuum Mechanics (CM)	1955	{2.2}, {2.4} etc.

[9.2] In the June of 1957, Noll finished a report for the Air Force Office of Scientific Research, entitled "On the Foundations of the Mechanics of Continuous Media". It wasn't accessible to the wider scientific public until April 22, 1993, and it isn't considered here.

[9.3] See {2.10}, p. 266.

[9.4] Ibid., p. 281.

[9.5] The axiom **N1** causes some difficulties for a description of contact interactions if two different points of a body can join each other.

[9.6] Regarding the mistake in the definition of the fit region of a continuous body in the axiom **N2** see Section 8.7.

[9.7] The axiom **N5** postulates that one can subdivide a continuum body into a finite or countable number of parts with masses. The mass of the body's point is undefined.

[9.8] See {2.10}, pp. 267–268.

[9.9] Ibid., p. 268. The mathematical theory behind the concept of motion was that of differentiable and continuous functions of the standard analysis. Noll's contribution to the axiomatization of the concept of motion is thus negligible. See, for example, the corresponding axioms in: G. Hamel, "Die Axiome der Mechanik", *Grundlagen der Mechanik: Mechanik der Punkte und starren Körper*, hrsg. v. R. Grammel, Springer-Verlag, Berlin, 1927, S. 1–42.

[9.10] See {2.10}, pp. 269–270.

[9.11] Ibid., p. 270.

[9.12] Ibid., pp. 272–273.

[9.13] Ibid., p. 280.

[9.14] Walter Noll gave the following characteristics of his axiomatic of continuum mechanics, presented earlier in {2.10}, in this Paris report: "In another communication {2.10} I presented an attempt at the axiomatization of general classical mechanics. The fundamental axioms were the laws of balance of actions and of moments and something that I called the principle of material objectivity which requires the invariance of the constitutive equations under arbitrary change of the frame of reference" ({2.22}, p. 48).

[9.15] See {2.22}, p. 48.

[9.16] He finally showed in 1965 that such a mathematical theory was the lattice theory (see {2.29R}, pp. 14–21). Walter Noll's works on contact interactions are described in Chapter 15.

[9.17] The axiom **N1Y** is equivalent to the axiom **N1X**. The axiom **N2Y** is a generalization of the axioms **N2X–N4X**.

[9.18] Walter Noll called a body $B \in U$ the "interior" of B. See {2.22}, p. 49.

[9.19] Ibid., pp. 48–49.

[9.20] Walter Noll remarked that "contact forces" existed only in continuum mechanics, while "distant forces" (forces de distance) could be attributed both to classical and continuum mechanics. See the axioms **N7Y–N10Y** in {2.22}, pp. 49–50.

[9.21] Ibid., pp. 50–51.

[9.22] Walter Noll remarked in {8.22}: "A function in a dynamic process is considered 'objective' if it changes in accordance with the laws of the change of frame of reference. Otherwise it is considered 'relative'" (p. 51).

[9.23] Ibid., pp. 52–53.

[9.24] Ibid., p. 53.

[9.25] Ibid., p. 54. Walter Noll gave also there an example of a postulate of the type **N16Y**.

[9.26] Ibid. Walter Noll gave also there an example of a postulate of the type **N17Y**.

[9.27] Ibid. Walter Noll also obtained there a general form of the postulate of energy.

[9.28] G. Hamel, "Die Axiome der Mechanik", *Grundlagen der Mechanik: Mechanik der Punkte und starren Körper*, hrsg. v. R. Grammel, Springer-Verlag, Berlin, 1927, S. 1–42.

[9.29] Ibid., p. 209.

[9.30] Ibid.

[9.31] See {2.9}, p. 200. All such deformations form a group, denoted by D.

[9.32] Ibid., p. 201.

[9.33] Ibid., pp. 209–210.

[9.34] Ibid., pp. 210–211. K_X^* is the set of all functions with values in K_X, whose domain is the negative real axis; K_X is the set of all local configurations of a particle X. Q^* is the set of all functions with values in Q, whose domain is the negative real axis. Q is the group of orthogonal linear transformations of V onto itself.

[9.35] Ibid., p. 211.

[9.36] Ibid., p. 212.

[9.37] Ibid.

[9.38] Ibid., p. 212.

[9.39] Ibid., p. 214. D^* is the set of all functions with values in D, whose domain is the negative real axis. D is a group of deformations.

[9.40] Here Ω^* is an element of D^*. See {2.9}, pp. 205, 214.

[9.41] Ibid., p. 216.

[9.42] Here L^* stands for the class of all functions with values in L, whose domain is the negative real axis. L is a linear group, i.e. a group of regular linear transformations. See {2.9}, pp. 217–218.

[9.43] Ibid., p. 218.

[9.44] Ibid.

[9.45] Ibid., pp. 218–219.

[9.46] Ibid., p. 219.

[9.47] Ibid.

[9.48] Ibid., pp. 221–225.

[9.49] C.A. Truesdell gave this characteristic of Mario Bunge in a letter to G. Oravas, dated February 19, 1969. Truesdell wrote there, in particular: "I agree that Bunge is a very good man. He is the only philosopher I know who has any contact with real science as it is practised. Unfortunately he publishes far too much and continually repeats himself with minor variations".

[9.50] M. Bunge, *Philosophy of Physics*, D. Reidel Publishing Co., Boston, 1973, pp. 159–161.

[9.51] Ibid., p. 133.

[9.52] M. Bunge, *Foundations of Physics*, Springer-Verlag, Berlin, 1967, p. 145.

[9.53] Ibid., pp. 145–151.

[9.54] Ibid., p. 145.

[9.55] See {2.37}, p. 1.

[9.56] Ibid., p. 2.

[9.57] Ibid., pp. 2, 49.

[9.58] Ibid., pp. 2–3.

[9.59] Ibid., pp. 5–8.

[9.60] Ibid., pp. 9–10. The motivations behind the definitions of the process and the body element were included by Walter Noll in Section 3 of {2.37}.

[9.61] Ibid., p. 11. See Walter Noll's motivations in Section 3 of {2.37}.

[9.62]   Ibid., p. 12.

[9.63]   Ibid., p. 13. Walter Noll wrote in {2.37} on the motivations behind the definition of material element: "A material element will be a body element endowed with additional structure that is designed to describe physical material properties. For definiteness we confine ourselves to mechanical material properties, although it is not difficult to include non-mechanical physical phenomena. . . . The principal feature of our definition is that it associates with the element a space of possible physical states. The underlying intuitive idea is that when a material element is given in a concrete physical situation, it is given in a definite state. The state determines everything about the element: its configuration, its stress, and, most importantly, the response of the element in every possible test. We imagine that a test consists of subjecting the element to a deformation process and measuring the stress at the end" (p. 12). It is clear from this explanation that Walter Noll showed himself in {2.37} as a cartesianist, for whom there could exist material elements without mass.

[9.64]   Ibid., p. 13.

[9.65]   Ibid., pp. 13–14.

[9.66]   Ibid., p. 14.

[9.67]   Ibid., p. 15.

[9.68]   Ibid., p. 17.

[9.69]   Ibid.

[9.70]   Ibid., pp. 18–20.

[9.71]   Ibid., p. 22.

[9.72]   Ibid.

[9.73]   Ibid., p. 25.

[9.74]   Ibid., pp. 25–48.

[9.75]   Ibid., p. 4.

[9.76]   Ibid., pp. 48–49. Walter Noll's interest in the general theory of physical systems can be dated back as far as to 1966, when he got to know H.D. Dombrowski and his works. Noll accomplished the proposed approach to thermodynamic processes in the expository paper {2.38}, where he built their exact axiomatical theory.

10 Rational Fluid Mechanics

[10.1] In 1957–1966, B.D. Coleman, through a great systematic effort, "learned the functional analysis required for the type of work described here".

[10.2] See {2.11}.

[10.3] Ibid., p. 289.

[10.4] Ibid., pp. 290–292.

[10.5] Ibid., p. 295.

[10.6] Ibid.

[10.7] Ibid., pp. 295–296.

[10.8] Ibid., pp. 296–297.

[10.9] Ibid., pp. 297–299.

[10.10] Ibid., pp. 299–300.

[10.11] Ibid., pp. 300–303.

[10.12] See {2.12}.

[10.13] Ibid., p. 1508.

[10.14] Ibid.

[10.15] Ibid.

[10.16] Ibid. Obviously, this scientific success of Walter Noll and B.D. Coleman played an important role in the future. R.S. Rivlin was deeply disturbed by it, and it caused, together with certain other reasons, the first priority dispute between Walter Noll and the British school of modern continuum mechanics.

[10.17] Ibid., pp. 1508–1509.

[10.18] Ibid., p. 1510.

[10.19] Ibid., p. 1512.

[10.20] See {2.16}, p. 672.

[10.21] Ibid.

[10.22] Walter Noll gave the following definition of a simple fluid in {2.16}: "The material at X is a *simple fluid* if the stress tensor $S(t)$ at X at time t depends on the history of the motion through an equation of the form

$$S(t) = \mathop{H}_{r=0}^{\infty} \left[C_t(t - r); \rho(t) \right].$$

Here, H is a functional which has for its argument the history of the right Cauchy-Green tensor relative to the configuration at time t and the density at time t. H is assumed to obey the following identity for all histories $C_t(t-r)$ and for all constant orthogonal tensors Q:

$$Q \overset{\infty}{\underset{r=0}{H}} [C_t(t-r); \rho(t)] Q^T = \overset{\infty}{\underset{r=0}{H}} [QC_t(t-r)Q^T; \rho(t)]$$

" (p. 679). As W. Noll and B.D. Coleman remarked in the introduction to {2.16}, "this definition was sufficiently general to include perfect fluids, Newtonian fluids, and viscoelastic fluids as special cases" (p. 672).

[10.23]   Ibid., p. 700. Walter Noll and B.D. Coleman gave a mathematically rigorous proof of (ii) at pages 702–707.

[10.24]   Ibid., pp. 701–702.

[10.25]   Ibid., p. 704.

[10.26]   Ibid., pp. 707–714.

[10.27]   See {2.19}, p. 841.

[10.28]   Ibid., p. 840. In (iii), the term "substantially stagnant motion" belonged to B.D. Coleman, who introduced it in 1961.

[10.29]   Ibid., p. 842.

[10.30]   Ibid.

[10.31]   Ibid., pp. 842–843.

[10.32]   Ibid., p. 843.

[10.33]   See {2.26}, pp. 530, 532.

[10.34]   Ibid., p. 531.

[10.35]   Ibid., pp. 531–532.

[10.36]   Ibid., pp. 532–540.

[10.37]   Ibid., pp. 540–551.

[10.38]   See {2.20}. B.D. Coleman's priority paper appeared as: "Kinematical Concepts with Applications in the Mechanics and Thermodynamics of Incompressible Viscoelastic Fluids", *Archive for Rational Mechanics and Analysis* **9** (1961), pp. 273–300.

[10.39]   See {2.20}, p. 99. $C_t^t(s) = C_t(t-s)$, where $C_t(\tau)$ is the relative right Cauchy-Green tensor.

[10.40]   Ibid., pp. 99–101.

[10.41]   Ibid., p. 102.

[10.42] Walter Noll gave the following definition of steady curvilinear flow: "We call steady curvilinear flow a motion whose velocity field $v(x)$ has the contravariant components $v^1 = 0$, $v^2 = u(X^1)$, $v^3 = w(X^1)$ in an orthogonal curvilinear coordinate system X^k whose metric components g^{kk} are constant along the curves $\xi = \xi(s)$ defined by

$$\xi^1 = X^1; \quad \xi^2 = X^2 + su(X^1);$$
$$\xi^3 = X^3 + sw(X3)"\quad \text{(ibid., p. 102)}.$$

[10.43] See {2.20}, pp. 104–105.
[10.44] See {2.23}, pp. 82–83.
[10.45] See the preface to {1.3}.
[10.46] See {1.3}, p. 7.
[10.47] The citation of A.C. Pipkin is taken from a brochure *Books Written or Edited by C. Truesdell*, Springer-Verlag, New York, 1989.
[10.48] Walter Noll and co-authors gave the following definition of viscometric flow in {1.3}, different from {2.20}: "Flows whose histories differ from $F(s) = 1 - sM$, where F stands for the relative deformation gradient, M is a tensor with the matrix

$$[M] = \begin{vmatrix} 0 & 0 & 0 \\ \chi & 0 & 0 \\ 0 & 0 & 0 \end{vmatrix}$$

$1(s) = 1$, $0 \le s \le \infty$ is called the rest history; it is the history of a volume element which has been in its present configuration at all times in the past only by changes of frame are called viscometric flows" (pp. 19, 22, 29). Thus the Cauchy-Green tensors were needed no more here. The viscometric flows included as special cases the curvilinear flows, flows through a channel, helical flows, flows between concentric cylinders, Couette flows, Poiseuille flows, cone and plate flows, torsional flows etc. They were all considered in this book.
[10.49] As the scientists remarked in the introduction to {1.3}, they didn't consider a theory of fading memory for viscometric flows, but it was, in principle, possible (pp. 8–9).

11   Rational Elasticity

[11.1]   See {2.17}, p. 239.

[11.2]   Ibid., p. 243.

[11.3]   Ibid., p. 244.

[11.4]   Ibid.

[11.5]   Ibid., pp. 242–245.

[11.6]   Ibid., p. 245.

[11.7]   Ibid.

[11.8]   Ibid., p. 246.

[11.9]   Ibid., pp. 246–247.

[11.10]   Ibid., pp. 248–249.

[11.11]   Ibid., p. 249. The attempt of Walter Noll and B.D. Coleman to rewrite viscoelasticity challenged the works of R.S. Rivlin and his collaborators, which were to become an internationally acknowledged standard. It influenced the attitude of R.S. Rivlin to Walter Noll and his works.

[11.12]   See {2.18}, p. 41.

[11.13]   See {2.28}, p. 95.

[11.14]   Ibid., p. 96.

[11.15]   Ibid.

[11.16]   Ibid.

[11.17]   Ibid.

[11.18]   Ibid., pp. 98–99.

[11.19]   Ibid., pp. 99–100.

[11.20]   Ibid., p. 100.

12   Rational Thermomechanics

[12.1]   The only exception was Walter Noll's report {2.28} on finite elasticity.

[12.2]   Letter of B.D. Coleman to Y.A. Ignatieff, dated April 13, 1990.

[12.3] See {2.13}, pp. 262, 265. Numerous signs testify that this paper was mostly written by B.D. Coleman. For example, the level of mathematics there was too low for Walter Noll.

[12.4] See {2.14}, p. 97. Walter Noll and B.D. Coleman excluded from this theory "the thermodynamics of chemical reactions, phase transitions, and capillarity". Noll compensated for this later, to a certain extent, in his paper {2.35}.

[12.5] Ibid., p. 102.

[12.6] Ibid.

[12.7] Ibid.

[12.8] Ibid.

[12.9] Ibid.

[12.10] Ibid., p. 103.

[12.11] Ibid.

[12.12] Ibid. Walter Noll and B.D. Coleman obtained the **TS8** as a corollary of the **TS7** and the polar decomposition theorem. However, it was easier to take it as an independent axiom.

[12.13] Ibid.

[12.14] Ibid.

[12.15] Ibid., pp. 103–104.

[12.16] Ibid., p. 104.

[12.17] Ibid., pp. 105–106.

[12.18] Ibid., p. 107. Walter Noll and B.D. Coleman called $\lambda^*(F, \eta)$ "free energy per unit mass of the local state (F, η) when under the action of the force temperature pair (s_r, θ)".

[12.19] Ibid., pp. 107–108.

[12.20] Ibid., pp. 98, 109.

[12.21] Ibid.

[12.22] Ibid., pp. 110–111. Walter Noll and B.D. Coleman remarked that the second fundamental postulate (**TS16**) guaranteed that the solution $\eta = \eta^{**}(F, \varepsilon)$ of the caloric equation of state was unique.

[12.23] Ibid., pp. 112–118.

[12.24] Ibid., p. 99.

[12.25] Walter Noll and B.D. Coleman wrote in {2.14}: "In the classical treatments of thermostatics ... the adjective *stable* is used in two senses. It is sometimes used as a modifier for the word *equilibrium*; i.e. one refers to 'status of stable equilibrium'; or it is used as a modifier for the word *state*; i.e. one refers to 'stable states'. In this paper we never use the word stable in the former sense. The theory which we develop here makes a careful distinction between *local* states, referring to a material point in a body, and *global* states, referring to the body as a whole. A local thermomechanic state is specified by giving the entropy density and the local configuration at a material point. A global thermomechanic state, on the other hand, is specified only when the entropy field and the complete configuration are specified for the entire body. We regard *thermal equilibrium* to be a property of *local states*. We consider just one type of thermal equilibrium. We define a state of thermal equilibrium as a local thermomechanic state which minimizes an appropriate potential rather than as a state at which a first variation vanishes. We regard *stability* as a property of only *global states*. We consider several types of stable states, defined as global thermomechanic states which minimize certain energy integrals subject to different constraints" (p. 98).

[12.26]   Ibid., p. 119.
[12.27]   Ibid., pp. 121–122.
[12.28]   Ibid., p. 122.
[12.29]   Ibid., p. 123.
[12.30]   Ibid., pp. 123–124.
[12.31]   Ibid., pp. 125–128.
[12.32]   "Current Contents", January 13, 1986, Nr. 2, p. 16.
[12.33]   Ibid.
[12.34]   See {2.15}, p. 357.
[12.35]   Ibid.
[12.36]   Ibid., p. 358.
[12.37]   Ibid., p. 359.
[12.38]   Ibid.
[12.39]   Ibid., pp. 358, 362.
[12.40]   Ibid., p. 363.
[12.41]   Ibid., p. 357.
[12.42]   Ibid., pp. 365–366.
[12.43]   Ibid., p. 368.
[12.44]   Ibid., p. 369.

[12.45]   Letter of B.D. Coleman to Y.A. Ignatieff, dated April 13, 1990.

[12.46]   See {2.21}, pp. 167–168.

[12.47]   Ibid., p. 169.

[12.48]   Ibid.

[12.49]   Ibid.

[12.50]   Ibid.

[12.51]   Ibid.

[12.52]   Ibid.

[12.53]   Ibid.

[12.54]   Ibid.

[12.55]   Ibid. At p. 170, Walter Noll and B.D. Coleman also gave two balance equations in a differential form, which demanded additional smoothness assumptions.

[12.56]   Ibid., p. 170. The sets of these four equations in a reduced form were obtained by Walter Noll and B.D. Coleman on the basis of the principle of material objectivity.

[12.57]   Ibid.

[12.58]   Ibid., p. 172.

[12.59]   Ibid., pp. 173–174. Walter Noll and B.D. Coleman re-marked on **TD13**: "The postulate . . . is a statement of the second law of thermodynamics within the framework presented here" (p. 174).

[12.60]   Ibid., pp. 174–176.

[12.61]   Ibid., p. 177.

[12.62]   See {2.25}, p. 87.

[12.63]   Ibid., p. 93.

[12.64]   Ibid.

[12.65]   Ibid., p. 95.

[12.66]   Ibid., p. 96.

[12.67]   Ibid.

[12.68]   Ibid., p. 97.

[12.69]   Ibid., p. 98.

[12.70]   Ibid., pp. 98–101.

[12.71]   Ibid., p. 103.

[12.72]   Ibid., pp. 104–106.

[12.73]   Ibid., pp. 107–109.

[12.74]   Ibid., pp. 109–111.

13 Neo-Classical Space-Time of Mechanics

[13.1] It is curious that Walter Noll's lectures in Bressanone {2.29} haven't been included in the list of references of his report at the Delaware Seminar {2.30}.

[13.2] See {2.30}, p. 29.

[13.3] Ibid.

[13.4] Walter Noll wrote in {2.30}: "The only language within which a non-classical structure can be unambiguously formulated is the language of axiomatic mathematics. The methods of modern mathematics make it possible to fabricate almost at will structures that can play the role of classical space-time" (p. 29).

[13.5] Walter Noll wrote in {2.30}: "Classical space-time with its absolute space has been most successful in those branches of mechanics in which inertia plays the central role. Such is not the case, however, in many of the branches of the mechanics of continuous media, where inertia is often of minor importance or sometimes even altogether neglected. In these branches of mechanics absolute place is an artificial and inappropriate concept. If it is used anyway, it is necessary to compensate for its arbitrariness by introducing a requirement of invariance, called the *principle of frame-indifference or objectivity.* ... The description of mechanics in terms of the neo-classical space-time, which has no room for an absolute space, shows clearly why this principle is needed" (pp. 29–30).

[13.6] See Chapter 4.

[13.7] See {2.29}, p. 7.

[13.8] See {2.30}, p. 30.

[13.9] See {2.29}, pp. 6–7; {2.30}, p. 30; {2.38}, p. 65; {2.40}, pp. 11–12.

[13.10] In {2.29}, Noll called the simultaneity-relation "the set of all pairs of simultaneous events $S = \{(a, b) \mid \tau(a, b) = 0\}$" (p. 7). Evidently, it turned out to be inconvenient, and he returned in {2.38} to the standard approach, used here.

[13.11] See {2.29}, p. 7; {2.30}, p. 30; {2.38}, p. 65; {2.40}, p. 12.

[13.12] See {2.29}, pp. 7–8; {2.30}, p. 31; {2.38}, pp. 65–66; {2.40}, pp. 11–12. The translation space of I_a is denoted in **D3** by V_a.

[13.13] The principle of objectivity was formulated by Walter Noll, for example, in {2.10}: "If a dynamical process is compatible with a constitutive assumption then all processes equivalent to it must also be compatible with this constitutive assumption. In other words, constitutive assumptions must be invariant under changes of frame" (pp. 280–281).

[13.14] See {2.40}, p. 14.

14 Inhomogeneities in Simple Bodies

[14.1] See {1.2}, pp. 88–89. Walter Noll remarked there that his theory of dislocations wasn't as general as some of the recognized contributions to it, since he made no use of couple-stresses. He wrote to justify himself: "That couple-stresses are really needed for an adequate description of the behavior of real solids has not yet been shown. A treatment of the polar-elastic media ... along the lines of Noll's theory of simple materials can surely be constructed" (p. 89).

[14.2] It sounds plausible to conclude that Noll's friends persuaded him to obstain from a publication of the preliminary manuscript. Between 1963 and the July of 1967, when the final version of the theory of dislocations was finished, Walter Noll worked tremendously to improve the text. He managed, in particular, to propose a coordinate- free type of all the necessary mathematical results from the contemporary differential geometry and to bring his theory in correspondence with the most important theories of "continuous distributions of dislocations" ({2.31}, p. 2).

[14.3] See {2.31}, {2.31R1}, {2.31R2} and {2.32}, {2.32R}.

[14.4] See {1.2}, pp. 90–91.

[14.5] See {1.22}, p. 91.

[14.6] Clifford A. Truesdell introduced the term "good theorems" in his expository book *Six Lectures on Modern Natural Philosophy*, Springer-Verlag, Berlin, 1966, p. 107.

[14.7] In the manuscript "The Decline of the Professor", Walter Noll described this boundary of his conceptual mathematics in the words that a conceptualization was to "unify and to simplify the results that had been found" by others (p. 2).

[14.8]   See {2.31}, p. 32, for references.

[14.9]   Ibid.

[14.10]   See {2.31}, p. 3.

[14.11]   Ibid., p. 2.

[14.12]   Ibid., p. 4.

[14.13]   Ibid., p. 6.

[14.14]   Ibid.

[14.15]   Ibid., p. 7.

[14.16]   Ibid.

[14.17]   In {2.32}, Noll gave an equivalent definition of material uniformity. He wrote: "We say that a body is materially uniform if it admits uniform references" (p. 243).

[14.18]   See {2.31}, p. 9.

[14.19]   See Section 8.1 for clarifications.

[14.20]   See {2.31}, pp. 9–10.

[14.21]   Ibid., pp. 10–18.

[14.22]   Ibid., p. 21.

[14.23]   See {2.31}, p. 24, or {2.32}, p. 244. L_B^r was a set of all tensor fields of class $\mathbf{C}^r$ on B ({2.31}, p. 11).

[14.24]   See Theorem 7 at p. 25 in {2.31}.

[14.25]   See {2.31}, p. 26, Theorem 8.

[14.26]   Ibid., p. 27, Theorem 9.

[14.27]   Ibid., Theorem 10.

[14.28]   See {2.31}, pp. 27–31; {2.32}, pp. 245–246.

15 Fit Regions and Contact Interactions

[15.1]   Epifanio Guido Virga was born on September 12, 1957, in Nicosia, Italy. He graduated in 1980 from the University of Pisa. His diploma was in mathematical physics. In 1983, he became an assistant professor in rational mechanics at the engineering school of the University of Pisa. Since 1991, Virga has been a professor of rational mechanics at the Department of Civil Engineering of the University of Pisa. He wrote in a letter to Y.A. Ignatieff, dated February 2, 1992: "I first met Walter Noll in Pisa during the winter of 1985. He was on leave from Carnegie-Mellon University and was offered a visiting professorship in Pisa for four months by Gianfranco Capriz, who is professor of mathematical physics at our university. He and his wife Resie arrived in January 1985. ... Walter came to the university quite regularly, at least three times a week. At that time I did not interact a great deal with him: I attended his lectures, I asked his advices, we occasionally had a chat, but that was almost all".

[15.2] Ibid. This paper of E.G. Virga was published as: "Metastable Equilibrium of Fluids with Surface Tension", *Quarterly of Applied Mathematics* **XLVI/2** (1988), pp. 217–228. He wrote there: "This paper deals with the effects of the surface tension on the equilibrium of fluids. It shows that, when a fluid with an incompressible fluid inclusion is put in a fluid environment kept at constant pressure, equilibrium configurations may arise whose character of stability is affected by the sign of pertubations of the environment pressure. Such configurations we refer to as metastable" (p. 217).

[15.3] Ibid. In the mentioned paper [8.263], E.G. Virga gave a definition of sets with finite perimeter as follows: "Let E be a Borel set of $\mathbf{R}^3$. Then the perimeter of E ... is the scalar

$$\underline{\mathrm{pm}}(E) = \underline{\sup} \left\{ \int_E \underline{\mathrm{div}}\, g \right\} ,$$

with the supremum taken on all functions $g : \mathbf{R}^3 \to \mathbf{R}^3$ of class $\mathbf{C}^1$ with compact support in $\mathbf{R}^3$, such that $|g(x)| \leq 1$, for every $x \in \mathbf{R}^3$" (p. 222).

[15.4] Ibid.

[15.5] See: M.E. Gurtin, W.O. Williams, W.P. Ziemer, "Geometric Measure Theory and the Axioms of Continuum Thermodynamics", *Archive for Rational Mechanics and Analysis* **92** (1986), 1–22. This paper has been cited by C.A. Truesdell as an example of the application of "sets of finite perimeter and the associated concepts of functions of bounded variation" to continuum mechanics in the second edition of his textbook *A First Course in Rational Continuum Mechanics*, Vol. 1, Academic Press, Boston, 1991, p. 88.

[15.6]   See {2.47}, p. 2.

[15.7]   Ibid. Walter Noll remarked: "The study of their class requires the use of the concepts of 'measure-theoretic' interior, closure, and boundary of a set. We (W. Noll and E.G. Virga) do not need these concepts here" (p. 2).

[15.8]   In {2.45}, Walter Noll wrote: "It has been clear for several years – to most people who have thought about the problem – that a suitable specification of N (class of fit regions) can probably be obtained by using the concept of a 'set with a locally finite perimeter' from geometric integration theory. However, I do not know of any investigator who has considered all of the conditions **N1–N6** in detail" (p. 20). In {2.47}, Noll decided the priority problem in favour of Italian science: "It seems that Banfi & Fabrizio ... were the first to propose, in the context of continuum physics, a class of sets involving the concept. However, their class does not meet the first of the requirements mentioned (Walter Noll's axioms of the material universe)" (p. 2).

[15.9]   C.A. Truesdell, *A First Course in Rational Continuum Mechanics*, Vol. 1, 2. Edition, Academic Press, Boston, 1991, pp. 88–92.

[15.10]   See {2.45}, p. 20.

[15.11]   Ibid.

[15.12]   In {2.45}, p. 20, Noll wrote on the problem of simplicity: "In any case it seems likely that an appropriate theory can be developed fairly easily. What seems more difficult is to cast such a theory into a form understandable to people who have not learned – and perhaps do not want to learn – the entire machinery of geometric integration theory as it now stands. It would be highly desirable to simplify matters enough so that the theory can be presented in a graduate introductory course on continuum mechanics" (p. 20). In {2.47}, he added to this the task to satisfy the specific needs of engineers: "It is also desirable that the class of fit regions include all that can possibly be imagined by an engineer but exclude those that can be dreamt up *only* by an ingenious mathematician" (p. 2).

[15.13]   C.A. Truesdell, *A First Course in Rational Continuum Mechanics*, Vol. 1, 2. Edition, Academic Press, Boston, 1991, pp. 88–92. Truesdell remarked there: "For the portions of the ... section (on fit regions) that differ from the text of the first edition I am deeply indebted to E. Virga and W.O. Williams" (p. 92).

[15.14]   See {2.47}, p. 1.

[15.15]   Ibid., p. 15.

[15.16]   See {2.45}, pp. 12–13.

[15.17]   See {2.47}, p. 16.

[15.18]   In {2.47}, Walter Noll and E.G. Virga called a boundary ∂S of a fit region S negligible if for every ε from $\mathbf{R}\backslash\{0\}$ one could find a finite collection of balls which covered ∂S, and the sum of those volumes did not exceed ε (p. 14).

[15.19]   Walter Noll and E.G. Virga remarked in {2.47} that it was possible to substitute condition (iv) by a weaker one: "(iv)′: D has a boundary of finite area-measure" (p. 16). However, the class of fit regions $\underline{Fr}'$, thus obtained, turned out to be smaller than $\underline{Fr}$. Simultaneously, the condition (iv)′ presupposed a knowledge of the theory of measure, which could make the understanding of fit regions more complex than it was. The authors called the formulation of (iv) as having been done "in terms of elementary mathematics".

[15.20]   See {2.47}, pp. 19–20, Example (3).

[15.21]   See {2.45}, pp. 20–21.

[15.22]   See, for example, the papers: (i) R.A. Toupin, "Elastic Materials with Couple-Stresses", *Archive for Rational Mechanics and Analysis* 11 (1962), 385–414; (ii) R.A. Toupin, "Theories of Elasticity with Couple-Stress", *Archive for Rational Mechanics and Analysis* 17 (1964), 85–112. In {2.48}, Walter Noll and E.G. Virga wrote about Toupin's work: "The idea of a 'line distribution of force over the edges of a body' seems to have occurred first to Toupin in 1962. . . . However, his treatment suffered from the same defect as most treatments of distributions of forces prior to 1960, namely these distributions were only implicit in formulas for resultant forces. The systems of forces giving rise to these results were not brought into the open" (p. 1).

[15.23]   Letter of E.G. Virga to Y.A. Ignatieff, dated February 2, 1992.

[15.24]   Ibid.

[15.25]   Ibid.

[15.26]   Ibid.

[15.27]   See {2.48}, p. 31.

[15.28]   Ibid., p. 1.

[15.29]   Ibid., p. 4.

[15.30]   Ibid., p. 5.

[15.31]   See {1.5}, Section 67, as Walter Noll remarked in {2.48}.

[15.32]   See {2.48}, p. 7.

[15.33]   Ibid., p. 8. This theorem was also introduced as a condition for the fit regions in {2.45}, pp. 12–13. See also {2.47}, p. 17.

[15.34]   See {2.48}, p. 8.

[15.35]   Ibid.

[15.36]   Ibid.

[15.37]   Ibid., p. 9.

[15.38]   Ibid.

[15.39]   Ibid., p. 10.

[15.40]   Ibid., pp. 10–11.

[15.41]   Ibid., pp. 11–13. The authors introduced also a differentiation between internal and peripheral wedges.

[15.42]   Ibid., pp. 13–14.

[15.43]   Ibid., p. 2.

[15.44]   Ibid., pp. 14–15.

[15.45]   Ibid., pp. 16–17.

[15.46]   Ibid., pp. 15–19.

[15.47]   Ibid., pp. 21–22.

[15.48]   Ibid., p. 24.

[15.49]   Ibid., p. 25.

[15.50]   Ibid., p. 26.

[15.51]   Ibid., pp. 28–30.

16   Foundations of Special Relativity

[16.1]   See {2.24}, p. 129.

[16.2]   The group G_c and its properties have been described in full by Hermann Minkowski in his report: "Raum und Zeit / Vortrag, gehalten auf der 80. Naturforscher-Versammlung zu Köln am 21.

September 1908", *Gesammelte Abhandlungen von H. Minkowski*, hrsg. v. D. Hilbert, Bd. 2, B.G. Teubner, Leipzig, 1911, 432–433.

[16.3]   For example, C.A. Truesdell in his famous textbook *A First Course in Rational Continuum Mechanics*, Vol. 1, 2. Edition, Academic Press, Boston, 1991, introduced a primitive concept of "event" which is very close to the Minkowskian "Weltpunkt". The totality of events built the "event-world" at Truesdell, and "die Mannigfaltigkeit aller denkbaren Wertsysteme x, y, z, t soll die *Welt* heissen" at Minkowski. The concept of Truesdell's "world line" is taken directly from the report of Minkowski (see Table N.2). So, it can be considered as established that Minkowskian space-time has actually influenced the fundamentals of modern mechanics. However, this influence didn't lead to a revision of the space-time of mechanics, as Minkowski had expected. For example, Truesdell distanced himself from the Minkowskian event-world, when he wrote in the textbook: "Various kinds of mechanics rest upon use of different event-worlds. For example, the event-world of relativity (i.e. of the Minkowskian chronometry) and for the mechanics of oriented materials differ from the event-world we use in this book" (p. 30).

[16.4]   See Section 8.5.

[16.5]   See, for example, the letter of Walter Noll to H. Giesekus, dated August 16, 1984, where Noll wrote in particular: "Kempers erroneously confuses *frames of reference* with *coordinate systems*".

[16.6]   J.W. Schutz, *Foundations of Special Relativity: Kinematic Axioms for Minkowski Space-Time*, Springer-Verlag, Berlin, 1973, 2.

[16.7]   Ibid.

[16.8]   P. Suppes, "Axioms for Relativistic Kinematics with or without Parity", *Proceedings of the International Symposium on the Axiomatic Method: with Special Reference to Geometry and Physics*, held at the University of California, Berkeley, on December 26, 1957–January 4, 1958, ed. by L. Henkin, P. Suppes, A. Tarski, North-Holland Publishing Company, Amsterdam, 1959, 291–307.

[16.9]   Ibid., p. 293.

[16.10]   Ibid., p. 291.

Table N.2. A comparison of the texts of C.A. Truesdell and H. Minkowski

C.A. Truesdell (1991)	H. Minkowski (1908)
A world-line is a curve (a curve is a piecewise differentiable, one-parameter family of events: $l = f(s)$, and s varies over some real interval) in W (the event-world) whose image in $E \times R$ (E is here a three-dimensional Euclidean point-space, R is a real line) associates one place to each time, so that we may represent a world-line as follows: $\lambda : S \to E$, S being an interval of R (p. 34).	Wir richten unsere Aufmerksamkeit auf den im Weltpunkt x, y, z, t vorhandenen substantiellen Punkt und stellen uns vor, wir sind imstande, diesen substantiellen Punkt zu jeder anderen Zeit wieder zu erkennen. Einem Zeitelement dt mögen die Änderungen dx, dy, dz der Raumkoordinaten dieses substantiellen Punktes entsprechen. Wir erhalten alsdann als Bild sozusagen für den ewigen Lebenslauf des substantiellen Punktes eine Kurve in der Welt, eine Weltlinie, deren Punkte sich auf den Parameter t von $-\infty$ bis $+\infty$ beziehen lassen (S. 432).

[16.11] A.A. Robb, *Geometry of Time and Space*, 2. Edition, University Press, Cambridge, 1936.

[16.12] Ibid., pp. 402–404.

[16.13] Ibid. Unfortunately, the list of works of John W. Schutz on the axiomatization of Minkowskian Chronometry ignores important German contributions. Their place in the field of special relativity should be evaluated in a special study. In 1938, a German mathematician, Hans Hermes, published a monograph entitled *Eine Axiomatisierung der allgemeinen Mechanik*, where he proposed "a set of axioms for the foundation of special relativity" (see review by B. Rosser in the *Journal of Symbolic Logic* **3** (1938) 119–120. It was reprinted in 1970 by the Verlag Dr. H.A. Gerstenberg in Hildesheim, Germany. In 1938, a German mathematician, Karl Schnell, published an axiomatic of special relativity in his doctoral dissertation *Eine Topologie der Zeit in logistischer Darstellung* (Philosophische und Naturwissenschaftliche Fakultät, Westfälische Wilhelms-Universität zu Münster). As further references to German contributions to the axiomatization of special relativity, one can take: (i) H. Reichenbach, *Axiomatik der relativistischen Raum-Zeit-Lehre*, Braunschweig: Fr. Vieweg & Sohn AG, Braunschweig,

1924; (ii) C. Carathéodory, "Zur Axiomatik der speziellen Relativitätstheorie", *Sitz. Ber. Preuss. Akad. Wiss.*, physik.-math. Klasse, 1924, 12. It is possible that there are even more axiomatics of special relativity, and it is an important task to compile a complete list of them in the future.

[16.14]   In Section 4 on "Temporal parity", Patrick Suppes posed the following problem: "Is it possible to find 'natural' axioms which fix a direction of time?" He remarked also that the axiomatization of special relativity, belonging to A.A. Robb, gave no answer to this (p. 307).

[16.15]   Walter Noll wrote at page 129 of {2.24}: "Chronometry is the science of the measurement of time intervals. Einstein's first paper was a critical study of time measurements. It was Minkowski, however, who made a geometric discipline out of Einstein's chronometry". This Noll's statement is partly false. It was not Minkowski who made the ideas of Einstein mathematical. The latter borrowed the ideas from Minkowski.

[16.16]   See {2.24}, pp. 129–130.

[16.17]   Ibid., p. 131.

[16.18]   Ibid., p. 132.

[16.19]   Ibid.

[16.20]   Ibid.

[16.21]   Ibid.

[16.22]   Ibid., p. 133.

[16.23]   Ibid.

[16.24]   Ibid., pp. 133–134.

[16.25]   Ibid., p. 134.

[16.26]   Ibid., p. 136.

[16.27]   Ibid.

[16.28]   Ibid., p. 137.

[16.29]   Ibid., p. 138.

[16.30]   Ibid., p. 138.

[16.31]   Ibid., p. 139.

[16.32]   Ibid., p. 140.

[16.33]   See [9.14].

[16.34]   See {2.24}, p. 141.

[16.35]   Ibid.

[16.36]   Ibid.

[16.37]   Ibid.

[16.38]   Ibid., p. 142.

[16.39]   See: G. Szekeres, "Kinematic Geometry: An Axiomatic System for Minkowski Space-Time", *J. Aust. Math. Soc.* VIII (1968), 134–160. Szekeres presented his own axiomatization as related to "solely kinematic notions such as time and motion". Contrary to Noll, he posed the demand that an axiomatic of special relativity should "make no assumptions about 3-space itself". Szekeres followed in his paper the lines devised by A.A. Robb (see [9.11]).

[16.40]   See: M. Bunge, *Foundations of Physics*, Springer-Verlag, Berlin, 1967. Bunge didn't mention other axiomatics of special relativity, but he introduced some philosophical boundaries for them, differentiating between kinematic interpretations, mechanism, operationalism, conventionalism, and formalism.

[16.41]   See: J.W. Schutz, *Foundations of Special Relativity: Kinematic Axioms for Minkowski Space-Time*, Springer-Verlag, Berlin, 1973, p. 2.

[16.42]   See {2.42}, pp. 371–372.

[16.43]   See {2.43}.

[16.44]   See {1.5}, p. 161.

[16.45]   The most recent monograph on Minkowskian Chronometry is: G.L. Naber, *The Geometry of Minkowski Space-Time: An Introduction to the Mathematics of the Special Theory of Relativity*, Springer-Verlag, Berlin, 1992. In this Naber has completely ignored the axiomatic systems for Minkowskian chronometry. The only exception was made for the A.A. Robb's result "on measuring proper spatial separation with clocks" (pp. VII, 62).

List of Works of Walter Noll

1. Books and Course Notes

1963 1.   *Tensor Analysis*, Baltimore, The Johns Hopkins University.

1965 2.   (Co-author C.A. Truesdell), *The Non-Linear Field Theories of Mechanics*, S. Flügge's Encyclopedia of Physics III/3, Berlin, Springer-Verlag.

1966 3.   (Co-authors B.D. Coleman and H. Markovitz), *Viscometric Flows of Non-Newtonian Fluids*, Berlin, Springer-Verlag.

1974 4.   *The Foundations of Mechanics and Thermodynamics: Selected Papers of W. Noll*, Berlin, Springer-Verlag.

1987 5.   *Finite-Dimensional Spaces: Algebra, Geometry, and Analysis*, Vol. 1, Dordrecht, Martinus Nijhoff Publishers.

1992 6.   (Co-author C.A. Truesdell), *The Non-Linear Field Theories of Mechanics*, S. Flügge's Encyclopedia of Physics III/3, 2. Edition, Berlin, Springer-Verlag.

2. Scientific Papers

1952 1.   (Co-author E. Mohr), Eine Bemerkung zur Schwarzschen Ungleichheit, *Mathematische Nachrichten* **7**, 55–59.

1955 2.   On the Continuity of the Solid and Fluid States, *Journal of Rational Mechanics and Analysis* **4**, 3–81.

2R.   On the Continuity of the Solid and Fluid States, *International Science Review Series* **8/II**, New York, Gordon & Breach, 1965.

3.   Die Herleitung der Grundgleichungen der Thermomechanik der Kontinua aus der statistischen Mechanik, *Journal of Rational Mechanics and Analysis* **4**, 627–646.

1957 4.   Verschiebungsfunkionen für Elastische Schwingungsprobleme, *Zeitschrift für Angewandte Mathematik und Mechanik* **37**, 81–87.

5.   (Co-author R. Finn), On the Uniqueness and Non-Existence of Stokes Flow, *Archive for Rational Mechanics and Analysis* **1**, 97–106.

6.   On the Rotation of an Incompressible Continuous Medium in Plane Motion, *Quarterly of Applied Mathematics* **15**, 317–319.

7.   On the Foundations of the Mechanics of Continuous Media, *Technical Report Nr. 17*, prepared under Contract Nr. AF18(600)-1138, Division File Nr. 1.17, AFOSR TN 57-352, AD 132 425, Carnegie Institute of Technology, Department of Mathematics, Pittsburgh 13, PA, Mathematics Division, Air Force Office of Scientific Research, June 1957.

1958 8.   (Co-author R.J. Duffin), On Exterior Boundary Value Problems in Linear Elasticity, *Archive for Rational Mechanics and Analysis* **2**, 191–196.

9.   A Mathematical Theory of the Mechanical Behaviour of Continuous Media, *Archive for Rational Mechanics and Analysis* **2**, 197–226.

9R1.   A Mathematical Theory of the Mechanical Behaviour of Continuous Media, *International Science Review Series* **8/II**, New York, Gordon & Breach, 1965.

9R2.   A Mathematical Theory of the Mechanical Behaviour of Continuous Media, *Continuum Theory of Inhomogeneities in Simple Bodies*, New York, Springer-Verlag, 1968.

9R3.   A Mathematical Theory of the Mechanical Behaviour of Continuous Media, *The Foundations of Mechanics and Thermodynamics: Selected Papers of W. Noll*, Berlin, Springer-Verlag, 1974.

1959 10.   The Foundations of Classical Mechanics in the Light of Recent Advances in Continuum Mechanics, *The Axiomatic Method with Special Reference to Geometry and Physics*, Amsterdam, North-Holland Publishing Co., 226–281.

10R.   The Foundations of Classical Mechanics in the Light of Recent Advances in Continuum Mechanics, *The Foundations of Mechanics and Thermodynamics: Selected Papers of W. Noll*, Berlin, Springer-Verlag, 1974.

11.   (Co-author B.D. Coleman), On Certain Steady Flows of General Fluids, *Archive for Rational Mechanics and Analysis* **3**, 289–303.

11R1.   (Co-author B.D. Coleman), On Certain Steady Flows of General Fluids, *International Science Review Series* **8/II**, New York, Gordon & Breach, 1965.

11R2.   (Co-author B.D. Coleman), On Certain Steady Flows of General Fluids, *The Foundations of Mechanics and Thermodynamics: Selected Papers of W. Noll*, Berlin, Springer-Verlag, 1974.

12.   (Co-author B.D. Coleman), Helical Flow of General Fluids, *Journal of Applied Physics* **10**, 1508–1514.

12R.   (Co-author B.D. Coleman), Helical Flow of General Fluids, *International Science Review Series* **8/II**, New York, Gordon & Breach, 1965.

13.   (Co-author B.D. Coleman), Conditions for Equilibrium at Negative Absolute Temperatures, *The Physical Review* **115**, 262–265.

14.   (Co-author B.D. Coleman), On the Thermostatics of Continuous Media, *Archive for Rational Mechanics and Analysis* **4**, 97–128.

14R1. (Co-author B.D. Coleman), On the Thermostatics of Continuous Media, *International Science Review Series* **8/III**, New York, Gordon & Breach, 1965.

14R2. (Co-author B.D. Coleman), On the Thermostatics of Continuous Media, *The Foundations of Mechanics and Thermodynamics: Selected Papers of W. Noll*, Berlin, Springer-Verlag, 1974.

1960 15. (Co-author B.D. Coleman), An Approximation Theorem for Functionals, with Applications in Continuum Mechanics, *Archive for Rational Mechanics and Analysis* **6**, 355–370.

15R1. (Co-author B.D. Coleman), An Approximation Theorem for Functionals, with Applications in Continuum Mechanics, *International Science Review Series* **8/II**, New York, Gordon & Breach, 1965.

15R2. (Co-author B.D. Coleman), An Approximation Theorem for Functionals, with Applications in Continuum Mechanics, *The Foundations of Mechanics and Thermodynamics: Selected Papers of W. Noll*, Berlin, Springer-Verlag, 1974.

1961 16. (Co-author B.D Coleman), Recent Results in the Continuum Theory of Viscoelastic Fluids, *Annals of the New York Academy of Sciences* **89**, 672–714.

17. (Co-author B.D. Coleman), Foundations of Linear Viscoelasticity, *Reviews of Modern Physics* **33**, 239–249.

17R1. (Co-author B.D. Coleman), Foundations of Linear Viscoelasticity, *International Science Review Series* **8/III**, New York, Gordon & Breach, 1965.

17R2. (Co-author B.D. Coleman), Foundations of Linear Viscoelasticity, *The Foundations of Mechanics and Thermodynamics: Selected Papers of W. Noll*, Berlin, Springer-Verlag, 1974.

18. (Co-author B.D. Coleman), Normal Stresses in Second-Order Viscoelasticity, *Transactions of the Society of Rheology* **5**, 41–46.

1962 19.   (Co-author B.D. Coleman), Steady Extension of In-
              compressible Simple Fluids, *The Physics of Fluids* **5**,
              840–843.

      20.   Motions with Constant Stretch History, *Archive for
              Rational Mechanics and Analysis* **11**, 97–105.

1963 21.   (Co-author B.D. Coleman), The Thermodynamics of
              Elastic Materials with Heat Conduction and Viscosity,
              Archive for Rational Mechanics and Analysis **13**, 167–
              178.

      21R.   (Co-author B.D. Coleman), The Thermodynamics of
              Elastic Materials with Heat Conduction and Viscosity,
              *The Foundations of Mechanics and Thermodynamics:
              Selected Papers of W. Noll*, Berlin, Springer-Verlag,
              1974.

      22.   La Mécanique Classique, Basée sur un Axiome d'Ob-
              jectivité, *La Methode Axiomatique dans les Mécani-
              ques Classiques et Nouvelles*, Paris, Gauthier-Villars,
              47–56.

      22R.   La Mécanique Classique, Basée sur un Axiome d'Ob-
              jectivité, *The Foundations of Mechanics and Ther-
              modynamics: Selected Papers of W. Noll*, Berlin,
              Springer-Verlag, 1974.

      23.   Modern Theories of Flow in Tubes, *Pulsatile Blood
              Flow*, New York, McGraw-Hill Book Company, 77–
              83.

1964 24.   Euclidean Geometry and Minkowskian Chronometry,
              American Mathematical Monthly **71**, 129–144.

      24R.   Euclidean Geometry and Minkowskian Chronometry,
              *The Foundations of Mechanics and Thermodynamics:
              Selected Papers of W. Noll*, Berlin, Springer-Verlag,
              1974.

      25.   (Co-author B.D. Coleman), Material Symmetry and
              Thermostatic Inequalities in Finite Elastic Deforma-
              tions, *Archive for Rational Mechanics and Analysis*
              15, 87–111.

      25R.   (Co-author B.D. Coleman), Material Symmetry and
              Thermostatic Inequalities in Finite Elastic Deforma-

tions, *The Foundations of Mechanics and Thermodynamics: Selected Papers of W. Noll*, Berlin, Springer-Verlag, 1974.

26. (Co-author B.D. Coleman), Simple Fluids with Fading Memory, *Proceedings of the International Symposium on Second-Order Effects in Elasticity, Plasticity and Fluid Dynamics*, Oxford, Pergamon Press, 530–552.

1965 27. Proof of the Maximality of the Orthogonal Group in the Unimodular Group, *Archive for Rational Mechanics and Analysis* **18**, 100–102.

27R. Proof of the Maximality of the Orthogonal Group in the Unimodular Group, *The Foundations of Mechanics and Thermodynamics: Selected Papers of W. Noll*, Berlin, Springer-Verlag, 1974.

28. The Equations of Finite Elasticity, *Symposium on Applications of Non-Linear Partial Differential Equations in Mathematical Physics (1964)*, Providence, American Mathematical Society, 93–101.

1966 29. The Foundations of Mechanics, *Non-Linear Continuum Theories (C.I.M.E. Lectures, 1965)*, Rome, Cremonesi, 159–200.

29R. The Foundations of Mechanics, *Report Series 66-2*, Department of Mathematics, Carnegie Institute of Technology, Pittsburgh.

1967 30. Space-Time Structures in Classical Mechanics, *Delaware Seminar in the Foundations of Physics*, Berlin, Springer-Verlag, 28–34.

30R. Space-Time Structures in Classical Mechanics, *The Foundations of Mechanics and Thermodynamics: Selected Papers of W. Noll*, Berlin, Springer-Verlag, 1974.

31. Materially Uniform Simple Bodies with Inhomogeneities, *Archive for Rational Mechanics and Analysis* **27**, 1–32.

31R1. Materially Uniform Simple Bodies with Inhomogeneities, *Research Report Nr. 67-26*, Department of

Mathematics, Carnegie Institute of Technology, August 1967.

31R2. Materially Uniform Simple Bodies with Inhomogeneities, *Continuum Theory of Inhomogeneities in Simple Bodies*, New York, Springer-Verlag, 1968.

31R3. Materially Uniform Simple Bodies with Inhomogeneities, *The Foundations of Mechanics and Thermodynamics: Selected Papers of W. Noll*, Berlin, Springer-Verlag, 1974.

1968 32. Inhomogeneities in Materially Uniform Simple Bodies, *IUTAM Symposium on Mechanics of Generalized Continua (1967)*, Berlin, Springer-Verlag, 239–246.

32R. Inhomogeneities in Materially Uniform Simple Bodies, *Continuum Theory of Inhomogeneities in Simple Bodies*, New York, Springer-Verlag, 1968.

33. Quasi-Invertibility in a Staircase Diagram, *Research Report Nr. 68-37*, Department of Mathematics, Carnegie Institute of Technology, Carnegie-Mellon University.

33R. Quasi-Invertibility in a Staircase Diagram, *Proceedings of the American Mathematical Society* **23**, 1–4 (1969).

1970 34. Representations of Certain Isotropic Tensor Functions, *Archiv der Mathematik* **21**, 87–90.

35. On Certain Convex Sets of Measures and on Phases of Reacting Mixtures, *Archive for Rational Mechanics and Analysis* **38**, 1–12.

1971 36. (Co-author H.D. Dombrowski), Annihilators of Linear Differential Operators, *Journal d'Analyse Mathématique (Jerusalem)*, 205–284.

1972 37. A New Mathematical Theory of Simple Materials, *Archive for Rational Mechanics and Analysis* **48**, 1–50.

37R. A New Mathematical Theory of Simple Materials, *The Foundations of Mechanics and Thermodynamics: Selected Papers of W. Noll*, Berlin, Springer-Verlag, 1974.

1973 38.   Lectures on the Foundations of Continuum Mechanics and Thermodynamics, *Archive for Rational Mechanics and Analysis* **52**, 62–92.

38R.   Lectures on the Foundations of Continuum Mechanics and Thermodynamics, *The Foundations of Mechanics and Thermodynamics: Selected Papers of W. Noll*, Berlin, Springer-Verlag, 1974.

1974 39.   The Role of the Second Law of Thermodynamics in Classical Continuum Physics, *Modern Developments in Thermodynamics*, Toronto, John Wiley & Sons, 117–119.

40.   On the Concept of the Symmetry Group of a Physical System, *Proceedings of the Symposium on Symmetry, Similarity, and Group Theoretic Methods in Mechanics*, Calgary, 83–99.

1976 41.   The Representation of Monotonous Processes by Exponentials, *Indiana University Mathematics Journal* **25**, 209–214.

1977 42.   (Co-author J.J. Schäffer), Orders, Gauge, and Distance in Faceless Linear Cones; with Examples Relevant to Continuum Mechanics and Relativity, *Archive for Rational Mechanics and Analysis* **66**, 345–377.

1978 43.   (Co-author J.J. Schäffer), Order-Isomorphisms in Affine Spaces, *Annali di Matematica Pura ed Applicata* **117**, 243–262.

44.   A General Framework for Problems in the Statics of Finite Elasticity, *Contemporary Development in Continuum Mechanics and Partial Differential Equations*, North-Holland Mathematics Studies **30**, 363–387.

1986 45.   Continuum Mechanics and Geometric Integration Theory, *Categories in Continuum Physics (1982)*, Berlin, Springer-Verlag, 17–29.

46.   (Co-author P. Podio-Guidugli), Discontinuous Displacements in Elasticity, *Finite Thermoelasticity (1985)*, Roma, Accademia Nazionale dei Lincei, 151–163.

1988 47. (Co-author E. Virga), Fit Regions and Functions of Bounded Variations, *Archive for Rational Mechanics and Analysis* **102**, 1–21.

1990 48. (Co-author E. Virga), On Edge Interactions and Surface Tension, *Archive for Rational Mechanics and Analysis* **111**, 1–31.

1992 49. Isocategories and Tensor Functions, *Research Report Nr. 92-147*, June, Department of Mathematics, Carnegie-Mellon University.

50. The Geometry of Contact, Separation, and Reformation of Continuous Bodies, *Research Report Nr. 92-NA-029*, August, Center for Nonlinear Analysis, Department of Mathematics, Carnegie-Mellon University.

1993 51. (Co-author R.E. Artz), Linearly Induced Mappings Between Cones of Quadratic Forms, *Manuscript*, Department of Mathematics, Carnegie-Mellon University.

Name Index

Abbott, E.A.   3, 6, 189, 190
Adams, W.W.   64
Adkins, J.E.   48, 162
Ahlberg, J.H.   57, 61
Alexander, R.T.   12
Ames, W.F.   61
Ances, W.F.   57
André, K.   18, 195
Angeloni, B.   205
Artz, R.E., Jr.   85
Auslander, J.   42, 64, 202

Beatty, M.   45
Beckenbach, E.F.   196
Bellman, R.   196
Benney, D.J.   57, 61
Bernardin, D.   88
Bernstein, B.   40f., 112f., 202
Birnbaum, Z.W.   61
Blackley, G.R.   57
Bliss, H.   208
Boardman, J.M.   93
Borel, A.   15
Bossert, W.H.   57, 61
Bouligand, G.   15f.
Bourbaki, N.   15, 28, 69, 181, 194
Bragg, L.E.   86, 208
Bressan, A.   208
Broughman, R.   208
Bunge, M.   47, 111f., 159, 187,
    203, 215, 234
Byrne, R.   26

Cain, R.   65

Capriz, G.   204, 227
Carathéodory, C.   159, 233
Carlson, P.A.   87
Carnegie, A.   37
Cartan, H.   15, 194
Chang, T.S.   204
Chatelet, A.C.   39
Chiou, S.-M.   86
Clark, L.L.   59
Coleman, B.D.   2, 40ff., 46, 52,
    55, 71f., 96, 119ff., 129ff., 137,
    141, 145, 147f., 152ff., 203f.,
    208, 217ff.
Cyert, R.M.   207

Day, W.A.   117, 204
De Benedetti, S.   68
Deckert, K.L.   61
Dedekind, R.   69
Deely, J.J.   52
De Giorgi, E.   169
Destouches, J.-L.   40
Dieudonné, J.   194
Dill, E.H.   53
Dillon, O.W., Jr.   46
DiPrima, R.C.   57, 61
Dombrowski, H.D.   51, 55, 64,
    91ff., 169, 205, 209ff., 216
Dorn, W.S.   57
Duffin, R.J.   38, 43, 202
Duhem, P.   137
Durelli, A.J.   54

Einstein A.   68, 181, 187, 210, 233

246   Name Index

Name Index

Subject Index

Springer-Verlag
and the Environment

We at Springer-Verlag firmly believe that an international science publisher has a special obligation to the environment, and our corporate policies consistently reflect this conviction.

We also expect our business partners – paper mills, printers, packaging manufacturers, etc. – to commit themselves to using environmentally friendly materials and production processes.

The paper in this book is made from low- or no-chlorine pulp and is acid free, in conformance with international standards for paper permanency.